WATER RESOURCE PLANNING, DEVELOPMENT AND MANAGEMENT

(NON) CONVENTIONAL APPROACHES APPLIED TO THE STUDY OF BRAZILIAN AQUIFERS

WATER RESOURCE PLANNING, DEVELOPMENT AND MANAGEMENT

Additional books in this series can be found on Nova's website under the Series tab.

Additional E-books in this series can be found on Nova's website under the E-book tab.

WATER RESOURCE PLANNING, DEVELOPMENT AND MANAGEMENT

(NON) CONVENTIONAL APPROACHES APPLIED TO THE STUDY OF BRAZILIAN AQUIFERS

DANIEL MARCOS BONOTTO
AND
EDSON GOMES DE OLIVEIRA

Nova Science Publishers, Inc.
New York

LIBRARY OF CONGRESS CATALOGING-IN-PUBLICATION DATA

Available upon Request
ISBN: 978-1-61668-770-0

Published by Nova Science Publishers, Inc. † New York

CONTENTS

PREFACE

The Earth's water is always in movement, and the water cycle (hydrologic cycle) describes the continuous movement of water on, above, and below the surface of the Earth. Water can change states among liquid, vapor, and ice at various places in the water cycle, with these processes happening in different time scales. Thus, the water cycle describes the processes that drive the movement of water throughout the hydrosphere, whilst a reservoir represents the water contained in different steps within the cycle. The largest reservoir is the collection of oceans, whereas the third is groundwater. The average time a water molecule will spend in a reservoir is the residence time, which usually ranges from 100 to 200 years in shallow groundwater but over 10,000 years in deep groundwater. In general, there is a relationship between the mineral composition of natural water and that of the solid minerals with which the water has been in contact. This relationship may be comparatively simple and uncomplicated, as in the case of an aquifer receiving direct recharge by rainfall and from which water is discharged without contacting any other aquifer or other water. Or the situation may be rendered very complex by influence of one or more interconnected aquifers of different composition, mixing of unlike waters, chemical reactions such as base exchange, adsorption of dissolved ions, and other factors like anthropogenic inputs. This book reports how different techniques can be utilized to investigate the water movement in different aquifer systems occurring in Brazil. Conventional methods to determine the hydraulic conductivity will be described, as well the use of the natural uranium isotopes ^{238}U and ^{234}U to investigate the groundwater flow and hydrogeochemical reactions taking place along it.

Chapter 1

INTRODUCTION

Most probably water has been in our solar system from the beginning and was formed by the thermonuclear fusion process that produced the elements of the periodic system and their compounds. It is generally accepted that most of the water in the hydrosphere has originated from degassing of the Earth's mantle by volcanic eruptions and surfacing lava (basalt) in the course of the 5 billion years of the Earth's existence (Vries, 2000).

The Earth's water is always in movement, and the water cycle, also known as the hydrologic cycle, describes the continuous movement of water on, above, and below the surface of the Earth. Water can change states among liquid, vapor, and ice at various places in the water cycle, with these processes happening in different time scales. The water cycle describes the processes that drive the movement of water throughout the hydrosphere, whilst a reservoir represents the water contained in different steps within the cycle. Thus, the hydrosphere can be characterized as a system of different reservoirs from which water, solutes and energy are exchanged by the hydrological cycle (Vries, 2000). On a large scale, this circulation is driven by thermal energy of solar radiation and by potential- and pressure energy produced by gravity. On a small scale, capillary and osmotic forces play a role in water transport in soil and plants, whereas geothermal energy produces thermo-mineral convection currents in deep aquifers (Vries, 2000).

The total amount of water in the hydrosphere is estimated at 1.4×10^9 km^3, of which 96% resides in the oceans (Table 1). The remaining 4% consists of freshwater, which exists and moves only by the virtue of the continuous distillation process that turns salt water into a freshwater by evaporation and subsequent condensation (Vries, 2000). Most of this freshwater (~2%) is locked up in ice caps, icebergs and glaciers. This small amount accounts for approximately 70-75% of all fresh water reserves on the planet (Baumgartner and Reichel, 1975; Berner and Berner, 1987). The water contained within all living organisms represents the smallest reservoir.

Table 1. Volume and fluxes of water and their turn-over time in the different compartments of the hydrosphere (Baumgartner and Reichel, 1975; Vries, 2000)

	Volume (10^3 km^3)	% of total freshwater	Flux (10^3 km^3/year)	Turn-over time (year)
SALT WATER				
Oceans	1,350,000		425	3,000[3]
FRESHWATER				
Ice	27,800	69.3	2.4	12,000[4]
Groundwater	8,000[1]	29.9	15	500[5]
Lakes	220[2]	0.55		
Soil moisture	70	0.18	90	0.8[6]
Atmosphere	15.5	0.038	496	0.03[7]
Reservoirs	5	0.013		
Rivers	2	0.005	40	0.05[8]
Biomass	2	0.005		
TOTAL	**40,114**	**100**		

1 < 5,000 m depth; 2 about 50% contains salt or brackish water; 3 flux is oceanic evaporation; 4 flux estimated from discharge; 5 flux estimated at 37% of total continental discharge (base flow); 6 flux estimated at 80% of continental rainfall; 7 flux is total rainfall = total evaporation; 8 flux is total discharge from continents.

The residence time of a reservoir within the hydrologic cycle is the average time a water molecule will spend in that reservoir. It is a measure of the mean age of the water in that reservoir, though some water will spend much less time than average and some much more. An estimate of the distribution of freshwater over the various global reservoirs and the turn-over time of the different reservoirs is given in Table 1. Most of the freshwater reserve actively involved in the hydrological cycle is found in the upper few kilometers of the subsurface. Groundwater can also spend over 10,000 years beneath Earth's surface before leaving, where old groundwater is called fossil water (Castany, 1982). Water stored in the soil remains there very briefly, because it is spread thinly across the Earth, and is readily lost by evaporation, transpiration, stream flow, or groundwater recharge. After evaporating, water remains in the atmosphere for about 10-11 days before condensing and falling to the Earth as precipitation.

The subsurface consists of an upper zone with soil water and a lower saturated part with groundwater, where they are separated by the groundwater table (Vries, 2000). The depth at which groundwater is encountered depends on topography, subsurface structure and climate. Total storage of groundwater in a layer is determined by the porosity, but the ease at which groundwater can flow depends on a combination of interconnection and size of the pores (Vries, 2000). Residence times of groundwater range between months for shallow drainage systems to thousands of years for deeper systems, and especially in arid areas with low recharge and large turnover time (Vries, 2000). Exceptional cases of fossil or residual groundwater flow can be found in extensive groundwater basins in desert areas (Vries, 2000).

During the water movement according to the hydrologic cycle, it may be verified water pollution, i.e. contamination of water by undesirable foreign matter. Pollution comes in many forms, some conventional and others toxic, thus, impacting the oceans, surface water, underground water, etc. National Primary Drinking Water Regulations (NPDWRs or primary standards) are legally enforceable standards that apply to public water systems. Primary standards protect public health by limiting the levels of contaminants in drinking water. The World Health Organization (WHO) set up some guidelines for drinking water quality which are the international reference point for standards setting and drinking-water safety (WHO, 2004). Several factors may be responsible by water pollution, and a list of possible causes related to groundwater systems is presented in Table 2.

Brazil owns the highest availability of renewable hydrological resources in world (8,233 km^3/year), however, it is the 26th country in terms of social water availability (48,314 m^3/inhab./year), as a consequence of the great unequal distribution of the hydrological resources in the country (57% is concentrated in northern and portions of northeastern and central western regions, where the population density is low) (Borghetti et al., 2004). This advanced degree of unequal water distribution has caused special attention to the groundwater resources due to problems related to the interaction between the society and the environment.

Many types of chemicals and substances have constituted important source of contaminants to groundwaters in Brazil and several techniques have been utilized to investigate the water movement in subsurface. This book reports the use of traditional and non-conventional tools to evaluate the groundwater transport in different aquifer systems occurring in Brazil.

Particular attention is directed for describing the potential use of the natural uranium isotopes ^{238}U and ^{234}U on providing ages for the groundwater residence time in the huge Guarani aquifer.

Table 2. The nature of groundwater pollution, according to Raymond (1988)

Type	Contamination Evidence	General Causes
Septic systems	• Waste water shows above ground. • Detection of excessive bacteria, and chemicals in well water tests.	• Poor installation and/or maintenance • Disposal of household chemicals, such as paint thinners, into system. • Overloading the system with a garbage disposal unit. • Use of septic tank cleaning additives. • Too many closely-spaced septic systems in a limited area.
Small disposal pits	• Petroleum odor in well water. • Other chemical odors. • Detection of chemical in well water tests	• Improper disposal of chemicals, oil, pesticides, other wastes and used containers. • Lack of disposal facilities for small amounts of hazardous wastes.
House and garden chemicals	• Detection of chemicals in well water tests.	• Improper use of storage. • Improper disposal in backyard, ditches, low ground, septic systems, and overgrown ditches.
Stream infiltration	• Odors in well water. • Other changes in well water quality. • Detection of excessive bacteria, or chemicals in well water tests.	• Polluted stream or lake water enters well.
Deicing salts	• Salty taste in well water. • High chloride level in well water tests.	• Runoff from salt storage piles and highways.
Landfills	• Observed leachate runoff. • Seepage into nearby surface water. • High levels of bacteria, organic chemicals, metals, nitrates in well water tests.	• Lack of control over leachate. • Permeable soil. • Leaky landfill cover. • No liner or liner failure. • Poor management. • Lack of control over hazardous wastes disposal

Type	Contamination Evidence	General Causes
Storage lagoons	• Spills. • Changes in color, taste, odor or well water. • Unhealthy or dead vegetation near lagoon. • Greener and more vigorous plant growth near lagoon. • Detection of excessive bacteria or chemicals in well water tests.	• Poor installation and maintenance. • Overflows. • Seepage. • Liner failure. • Structural collapse. • Location in sensitive groundwater area.
Underground storage tanks	• Petroleum odor in wells or basements. • Tank inventory losses. • Spills. • Detection of leaks.	• Corroded tanks. • Poor installation and/or maintenance. • No testing for tank leaks. • Poor inventory control. • No leak backup containment. • Deterioration of abandoned tanks.
Fertilizers	• High nitrate level in well water tests.	• Over fertilization. • Ill-timed application.
Land application of sludges and waste water	• High bacteria, nitrate levels in well water tests.	• Improper application methods. • Inappropriate soils for application.
Pesticides	• Detection of pesticides in well water tests. • Ill effects on animals drinking from nearby wells, springs or surface water. • Ill effects on plants watered with nearby well water. • Ill effects on aquatic life.	• Excessive or ill-timed application. • Improper storage. • Leaching through soil. • Improper disposal of excess pesticides and rinse water.
Hazardous materials	• Spills. • Detection of chemical solvents, metals, nitrates, or other chemicals in well water tests.	• Improper storage, handling, use, and disposal. • Spills. • Leaks.
Transport and transfer spills	• Spills.	• Accidents. • Inadequate maintenance. • Lack of proper training. • Inadequate provisions for spill containment.
Pipelines	• Odors in basement or well water. • Pipeline flow losses. • Detection of pipeline leaks through inspections. • Detection of excessive bacteria, or chemicals in well water tests.	• Corrosion. • Defective welds. • Vibrations from heavy vehicles. • Displacement by tree roots. • Poor installation. • Poor maintenance and inspection.

Table 2. (Continued).

Type	Contamination Evidence	General Causes
Wells	• Detection of high bacteria levels in well water tests. • Well water turbidity. • Detection of other contaminants in well water tests.	• No well casing or leaky casing. • Well cover not watertight. • Open abandoned wells. • Groundwater movement from contaminated to uncontaminated wells.
Inactive mining sites	• (Potential) Dumping of wastes in inactive mining pits.	• Rapid infiltration of contaminants due to loss of topsoil filtering capacity.
Animal lots	• High bacteria or nitrate levels in well water tests.	• High animal density. • Shallow depth to water table. • Poor lot drainage. • Failure to regularly clean lot.
Urban runoff	• Detection of chemicals, metals, nitrates, petroleum, or other chemicals in surface water or well water tests.	• Spills. • Random waste disposal. • Abandoned commercial/industrial sites. • Motor vehicle emissions. • Fires.
Construction excavation	• Spills. • Changes in color, taste, odor, or turbidity of water in nearby wells.	• Fuel and chemical spills. • Road dust-control chemical runoff. • Excessive and/or improper use of chemicals.
Cemeteries and excavation	• Detection of high bacteria levels in nearby water well tests.	• High water table.
Natural substances	• Bad taste or odor in well water. • Stains on water fixtures. • Detection of unacceptable levels of substances in well water tests.	• Natural soil or rock conditions.

Chapter 2

Traditional and Non-Conventional Tools for Investigating the Groundwater Movement

The basis for investigating the groundwater movement provides from the initial studies held by the french hydraulic engineer Henri Darcy, while studying the fountains in the city of Dijon, France, around 1856. He investigated the flux of water through horizontal sandy beds that would be utilized for filtering the water and pointed out (Darcy, 1856): "I have tried by precise experiments determine the water flux law through filters… The experiments positively demonstrated that the water volume flowing through a sandy bed is directly proportional to pressure and inversely related to the bed thickness. Thus, naming s the surface area of a filter, K the coefficient depending of the sandy bed nature and thickness (L), $P - H_0$ the pressure under the sandy bed and $P + H$ the atmospheric pressure added to the water column over the filter, the flux in such conditions is $Q = (Ks/L)\ (H+L+H_0)$ that reduces to $Q = (Ks/L)\ (H+L)$ when $H_0 = 0$, i.e. when the pressure under the filter is equal to the atmosphere weight."

The assertive that the flux through a porous medium is directly proportional to the load loss and inversely related to its thickness has been universally known as Darcy's Law, which supports our present knowledge of the groundwater flow. The analysis and solution of problems related to the groundwater movement and wells hydraulics started from the pioneering studies of Henri Darcy. Another way to express the Darcy's Law is according to the formula (Todd, 1959; de Marsily, 1986): $Q = KA\ \Delta h/L$, where: A is the area of the cross-section of the sandy formation, Δh the difference in hydraulic head in the water between the top and the bottom of the sandy formation, K a constant that depends on the porous medium, called hydraulic conductivity in hydrogeology, or sometimes coefficient of permeability, and L the thickness of the sandy formation.

The filtration velocity U is expressed on dividing both sides of the previous formula by A, which represents a fictitious velocity of the fluid at the outlet of the formation, i.e. considers the entire section to be open to the flow (de Marsily, 1986): $U = Q/A$.

If $i = \Delta h/L$ is called hydraulic gradient that denotes the difference in hydraulic head per unit length of porous medium traveled by the flow, then, the Darcy's Law has the simplest expression (de Marsily, 1986): $U = Ki$.

A wide range of modifications of the original formulation of the Darcy's Law has been realized along the time, taking in account the specificity of the problems focused either in field or laboratorial studies. One possible version will be later described in this book for field studies held at Analândia municipality, São Paulo State, Brazil.

Soddy, a chemist at Glasgow University, in 1913, used the word isotope initially with regard to radioactive (unstable) atoms for explaining some results from his investigations that several elements occupied the same place in the periodic table (Greek isos = "equal", tópos = "site, place"). However, as part of his exploration into the composition of canal rays, J. J. Thomson channeled a stream of ionized neon through a magnetic and an electric field and measured its deflection by placing a photographic plate in its path. Thomson observed two patches of light on the photographic plate, which suggested two different parabolas of deflection and was the first observation of different stable isotopes for an element.

Since those initial studies, several applications have existed that capitalize on properties of the various isotopes of a given element. For instance, in biogenic substances, significant variations of isotopes of C, N and O can occur, whose analysis has a wide range of applications, such as the detection of adulteration of food products (Bigeleisen and Wolfsberg, 1958). Other applications include the identification of certain meteorites as having originated on Mars, the isotopic labeling (the use of unusual isotopes as tracers or markers in chemical reactions) and the radiometric dating (the evaluation of the time that has elapsed since a certain amount of isotope has existed, using the known half-life of an unstable element) (Sitte, 1967; Mook and Waterbolk, 1985; Faure, 1986).

The use of isotopic methods in groundwater investigations also gained widespread acceptance among hydrogeology professionals. Well-established techniques such as using stable isotopes of hydrogen and oxygen as markers of water source have been applied in water resource investigations for several decades. Isotope methods are powerful tools when applied to the intractable problems of source attribution for the most common groundwater contaminants, including nitrate, VOCs, and perchlorate (Fritz and Fontes, 1980; Mook, 2000). Radioactive isotopes can be used to determine groundwater age, or the time since water entered the saturated zone, which has important implications for delineating groundwater or contaminant flow pathways, identifying recharge areas, and providing a measure of aquifer vulnerability (Ivanovich and Harmon, 1982; Mook, 2000).

Two fundamental areas of application of the environmental isotopes have mainly occurred: in groundwater resources and in contaminant hydrogeology. Several studies were held considering the oxygen-18 and deuterium partitioning through the hydrological cycle, also including recharge origin and resource renewability. Approaches to answering these questions have focused ^{18}O and D during precipitation, evaporation, recharge, groundwater/surface water interaction, and mixing (Fritz and Fontes, 1980; Mook, 2000). The use of tritium also have occurred for constraining the residence time of recent recharge in studies of groundwater recharge and circulation rates, as well methods have been developed for modeling radiocarbon ages of dissolved inorganic carbon in older groundwater (Fritz and Fontes, 1980; Mook, 2000).

Groundwater quality studies focusing the understanding where contaminants come from and the transformations that they may experience in the subsurface geochemical environments have also been realized. Such investigations have included ^{13}C, ^{2}H, ^{37}Cl, ^{15}N and ^{34}S to trace the origin of contaminants and attenuation processes active in the aquifer, as well covering the groundwater contamination due to agricultural (nitrate), industrial (LNAPLs and DNAPLs), urban, and mining activities (Fritz and Fontes, 1980; Mook, 2000).

The natural uranium isotopes ^{238}U and ^{234}U have also been extensively utilized in hydrogeological investigations. The $^{234}U/^{238}U$ activity ratio (AR) is unity in the bulk of systems older than 1 Ma and that are closed for U, however, rock/soil-water interactions imply on disequilibria conditions that frequently results in ARs > 1 for dissolved uranium (Osmond and Cowart, 1976; Ivanovich and Harmon, 1982). Hydrological studies of groundwaters have utilized U contents and ARs data with the purpose of dating, among others (Andrews and Kay 1978, 1983; Andrews et al., 1982). One evaluation of the possibility of using these isotopes in hydrogeochemical investigations held at Guarani aquifer, Brazil, will also be later described in this book.

Chapter 3

HYDRAULIC CONDUCTIVITY AT ANALÂNDIA MUNICIPALITY

GEOLOGICAL ASPECTS OF THE AREA

The evaluation of the hydraulic conductivity was performed by Oliveira and Bonotto (2002) at two selected sites located in Analândia municipality, São Paulo State, Brazil (Figure 1). It is circa 230 km distant from São Paulo city, the capital of São Paulo State, and may be reached by the following roads: SP-330 (Anhanguera), SP-310 (Washington Luiz) and SP-225 (Brotas-Piraçununga). The borders of Analândia municipality are: Itirapina (west), São Carlos (north), Piraçununga (east) and Corumbataí (south).

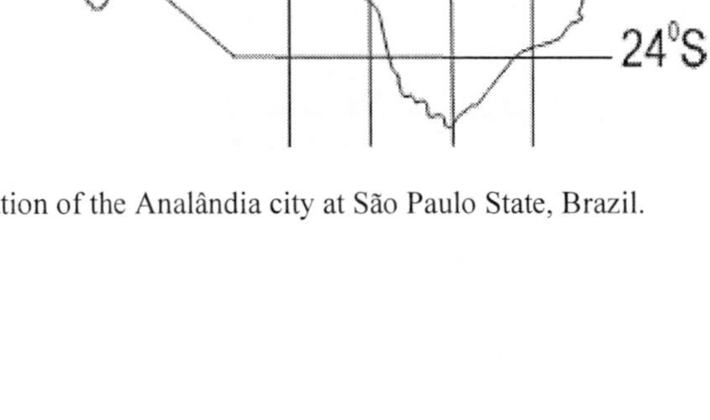

Figure 1. Location of the Analândia city at São Paulo State, Brazil.

The *Cwa* climate (Köppen classification) characterizes Analândia city, i.e. the winter is dry and the summer is hot and rainy. It is also named "tropical climate of altitude", whose pluviometric index varies between 1,100 and 1,700 mm. The dry season occurs between April and September, where July is often the driest month (the total monthly rainfall is commonly lower than 30 mm). The annual average rainfall in the area is 1,340 mm (Liazi et al., 1988), with the highest precipitation occurring in January and February. The average air temperature in the hottest months oscillates between 22°C and 24°C, whereas the average temperature in the coldest month is lower than 18°C and higher than 15°C. According to the Gaussen classification, the climate of the area is 4CTh-type (tropical hot with accentuated drought), which corresponds to a short hot season (duration of three/four months) that exhibits a xerotermic index varying between 400 and 100.

The two areas investigated (North Area and South Area) are geologically situated in the Paraná sedimentary basin. The basin is located in the South American continent, between parallels 10°-20° southern latitude and meridians 47°-64° western longitude. Its surface area corresponds to 1.6×10^6 km^2, comprising southern Brazil (1×10^6 km^2 in the states of Mato Grosso, Mato Grosso do Sul, Goiás, Minas Gerais, São Paulo, Paraná, Santa Catarina and Rio Grande do Sul), eastern Paraguay (0.1×10^6 km^2), NW Uruguay (0.1×10^6 km^2) and the northeastern extreme corner of Argentina (0.4×10^6 km^2) (Araújo et al., 1995). The altitudes in Paraná basin vary from 100 to 1800 m, the annual average rainfall corresponds to 100-240 cm, the soils are mainly lateritic, and the vegetation is often represented by grass-covered fields used for pasture and *cerrado* fields.

The Paraná basin is intercratonic, where the sedimentary sequence covers since the Silurian-Devonian up to the Cretaceous periods. The sequence is almost undisturbed, with gentle dips towards the center of the basin. Local faults may have served as channels for the extruding basalt flows of Jurassic-Cretaceous age (Serra Geral Formation), which cover about 1×10^6 km^2 of the basin. Renne et al. (1992) utilized the ^{40}Ar-^{39}Ar technique to estimate volcanism ages between 127 and 137 Ma, and Turner et al. (1994) suggested that the duration of the volcanism corresponded to 10 Ma, occurring at a rate of 0.1 km^{3}yr^{-1}.

The major stratigraphic units occurring at the Paraná basin are (IPT, 1981): the Tubarão Group comprising the Itararé Subgroup (sandstones, conglomerates, diamictites, tillites, siltstones, shales and rythmites) and Tatuí Formation (siltstones, shales, silex and sandstones with local concretions); the

Passa Dois Group comprising the Irati Formation (siltstones, mudstones, black betuminous shales and limestones) and Corumbataí Formation (mudstones, shales and siltstones); the São Bento Group comprising the Pirambóia Formation (sandstones, shales and muddy sandstones), Botucatu Formation (sandstones and muddy sandstones), Serra Geral Formation (basalts and diabases) and related basic intrusives; different types of Cenozoic covers like the recent deposits, terrace sediments and the Rio Claro Formation (sandstones, conglomerate sandstones and muddy sandstones).

The North Area was located at Santo André farm and belonged to CRS-*Mineração Indústria e Comércio Ltda.*, being situated about 5 km from Analândia city. The South Area was located at São João farm and belonged to *Sibelco Mineração Ltda.*, being situated approximately 7 km from Analândia city. Several stratigraphic units of the Paraná sedimentary basin cropp-out in the region, where the main features of the geology at the studied areas are similar and whose lithologies include the Lower and Upper units from Pirambóia Formation, the Botucatu Formation and the weathered cover developed over Pirambóia Formation (red-yellow latosols). The Lower unit from Pirambóia Formation consists of a consolidated whitish sandstone (after washing) with well-preserved sedimentary structures, whose thickness is more than 80 m, the retention after grinding is greater on the 0.125 mm-size sieve, and the Fe oxide mineral content is low. The occurrence of small clay lenses of variable lateral size is very common in this unit. The Upper unit from Pirambóia Formation is a light pink saprolite characterized by thin sedimentary structures and a very brittle sandstone located above the water table, whose concentration of Fe oxide is also low and the dominant particle-size after milling is in the range 0.125-0.25 mm. A 1-30 cm thick stone line profile (Lecomte, 1988) occurring as an irregular replica of the topography divides this unit and the above located mantle of weathering, which is constituted by tabular blocks of sandstones cemented by black colored limonite, dark red ferruginous concretions, small tablets of limonite cementing sandstone, SiO_2 pebbles, weathered yellowish sandstone and major concentration of heavy minerals (hematite, magnetite and ilmenite). The Botucatu Formation is represented in the area by weathered sandstones with high content of Fe oxide and particle-size predominantly in the range 0.5-1 mm. The weathered cover developed over Pirambóia Formation includes: peat soil, creep soil, and soil above the stone line. The peat soil is light grey-black colored, and occurs from the point of the discharge of the water table up to the contour level corresponding to that of the local streams. The creep soil is very clayey (12-20%), occurs on the steep ground, being characterized by a red-brownish colour, values of Fe oxide content of about 5000 ppm, and presence of limonite concretions between 2 and 6 m depth. The soil above the stone line is yellow-orangish colored, its thickness may attain 18 m, occurs where the ground slope is smooth, has a clay content varying between 7 and 18%, and a higher Fe oxide content near the surface due to the laterization, which diminishes with increasing depth, rising again at the stone line position. A simplified geological map of the area is shown in Figure 2, with indication of the South Area.

Figure 2. Simplified geological map at the South Area (Research Area) in Analândia city, São Paulo State, Brazil. Modified from IPT (1981), IG (1984) and Tandel (1993).

Hydrological Aspects of the Areas Investigated

In terms of the hydrological context, the areas investigated are situated in the Corumbataí River hydrographic basin that is a sub-basin of the Piracicaba River basin (Oliveira and Bonotto, 2002). More specifically, the two areas are located in the following micro basins: Veado Stream (North Area) and Taipas Creek (South Area) (Figure 3). The Veado Stream (Current) flows in the E-W direction along an opened V-valley whose northern and southern portions exhibit a general gradient that may reach a value of up to 15% and whose total longitudinal length is about 3,000 m from its rising up to the confluence with Corumbataí River. The average discharge of this current is 24.9 L/s, reducing approximately 10% in the dry periods. Because it is a current of short length and since it starts flowing in the higher topographic levels from the local Pirambóia Formation, the general gradient of the channel corresponds to 2.3%, whereas the surface area of this micro-basin is about 2.26 km^2. Such drainage channel is mainly supplied by rainfall and, secondarily, by groundwater discharging from risings and springs. The available data providing from monitoring wells and outcropping water level indicate that the groundwater flow in Pirambóia Formation at North Area has a general direction from SE to NW, with piezometric gradients equivalent to 0.07 (7% or 4°). The local sedimentary package is thin (maximum of 18 meters from the datum) and the soil is very porous, permeable, friable and with low compaction, thus, these factors coupled to the narrowing of the piezometric contour curves contribute for the low hydraulic gradient verified in the area if compared with the permeability (Oliveira and Bonotto, 2002).

The major contribution for surface waters in South Area provides from Taipas Creek that flows in the NE-SW direction (Figure 3). The average discharge of this stream is 337.49 L/s, also reducing approximately 10% in the dry periods. The general gradient of this stream from its origin up to the confluence with Corumbataí River corresponds to 2%, despite lower values (~1.3%) often occur where sand mining activities have taken place (Oliveira and Bonotto, 2002). This area is located in the northern part of the stream, in which the groundwater flow lines exhibit some peculiar characteristics, for instance: a) the flow net contains more accentuated gradients, varying between 0.1 (10 % or 6°) and 0.15 (15% or 8°); b) the mean direction of the flow vectors varies from N-S to NW-SE, as expected, but there are some distortions in the flow lines directions associated to some monitoring wells due to two major factors. The first refers to the existence of a fault line in this place, responsible by the generation of a more permeable zone in it, which caused a lowering of the water level in monitoring wells and, as a consequence, modification of the local network flow. The second factor is related to the sedimentary nature, i.e. the flow net occurs in the Lower Unit of the Pirambóia sandstone, which reveals a more accentuated control of their directions that are not completely subordinated to the hydraulic gradient but are linked to the less permeable and more compact character of this unit. This aspect is reinforced by the lower inflection of the flow lines along the free surface of the phreatic level (in the stream) as well by their more accentuated gradient (Oliveira and Bonotto, 2002).

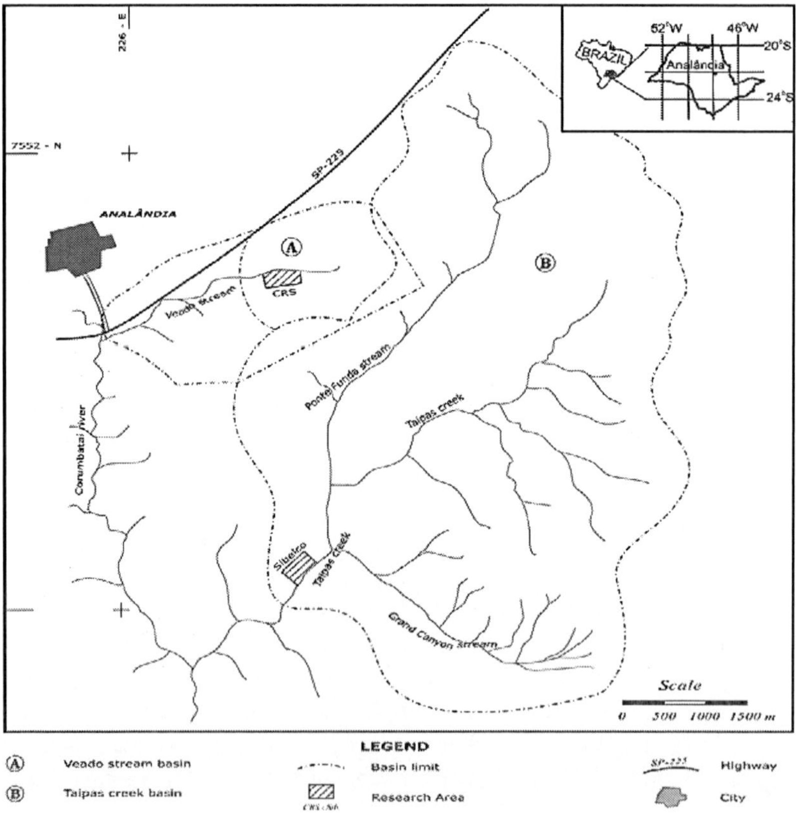

Figure 3. Simplified hydrological map at the areas investigated in Analândia city, São Paulo State, Brazil. According to Oliveira and Bonotto (2002).

Veado Stream and Taipas Creek Sub-Basins

The main characteristics of the Veado Stream and Taipas Creek sub-basins are summarized in Table 3. The parameters expressed in Table 3 were defined elsewhere (Pinto et al., 1973; Esteves, 1988; Liazi et al., 1988), as follows:

- **UGRHI:** the code means Unit for the Management of Hydrological Resources, and identifies all hydrographic basins in São Paulo State, Brazil (Liazi et al., 1988). It corresponds to 14 in the studied area.
- **Axial length:** it corresponds to the linear distance of the main axis of the basin that is measured from the point of discharge of effluent up to the more distant of the basin.
- **Average width:** it corresponds to the average distance of the transverse to the axial length of the basin.
- **Area:** it corresponds to the estimated area of the basin.
- **Perimeter:** it corresponds to the linear extension limiting the basin.
- **Average ground slope:** it corresponds to the average ground slope of the main axis of the drainage that is measured in percentage considering the more elevated and depressed points.
- **Total length of the drainage:** it corresponds to the summation of the linear lengths of all drainages occurring in the basin.
- **Order of the drainage:** it corresponds to the ramification degree of the drainage.
- **Density of the drainage:** this parameter compares the total length of the drainage (L, in km), with the basin area (A, in km^2), i.e. $D_d = L/A$. Higher values imply on higher amount of water in the basin, facilitating the flooding and, consequently, not favoring the flow. Typical values are: $D_d = 3$ (dense drainage), $D_d = 1\text{-}2$ (average drainage), and $D_d < 1$ (poor drainage).
- **Surface flow coefficient:** this parameter evaluates the average distance of the rainwater flow over the soil of the basin, i.e. it is the reciprocal of the density of the drainage. It may be estimated by the equation $E_s = A/4L$, where, E_s = surface flow, A = basin area, and L = total length of the drainage.
- **Shape factor:** this parameter focuses the basin shape, relating the basin area (A) and the total length of the drainage (L) by the equation $K_f = A/L^2$. Higher K_f values (higher than 2) tend to facilitate flooding.

- **Gravelius coefficient:** this parameter relates the perimeter of the basin (P) with a circle possessing an area equivalent to that of the basin (A), according to the equation $K_c = 0.28\ P/(A)^{1/2}$, where 0.28 is a constant. The resulting value is dimensionless, which generally exceeds 1. Values close to 1 imply that the basin shape practically corresponds to a circle, favoring the occurrence of flooding. In this case, the North Area tends to flood more easily than the South Area, according to K_c values.

Table 3. Parameters related to the sub-basins investigated

Parameter	Sub-basin	
Name	North- CRS	South- Sibelco
Main drainage	Veado stream	Taipas creek
Dominant basin	Corumbataí river	Corumbataí river
UGRHI	14	14
Axial length (km)	2.1	6.2
Average width (km)	15	5
Área (km^2)	2.3	30.6
Perimeter (km)	5.4	27.4
Average ground slope (%)	2.3	2
Total length of the drainage (km)	1.35	41.55
Order of the drainage	1st	4th
Density of the drainage (km/km^2)	0.59	1.35
Surface flow coefficient (km)	0.42	0.18
Shape factor - K_f - (km^2/km)	0.51	0.78
Gravelius coefficient - K_c - (dimentionless)	1.02	1.38
Mean annual discharge – Q_m– (L/s)	24.9	337.5
Minimum annual discharge for consecutive 7 days - $Q_{7,10}$ - (L/s)	5.56	75.43

Table 4. Parameters related to hydrographic basins in São Paulo State, Brazil. According to Liazi et al. (1988)

Region	Q_m		X_t						A/B values	
	a	b	Returning Period T (years)						A	B
			10	15	20	25	50	100		
A	-22.14	0.0292	0.708	0.674	0.655	0.641	0.607	0.581	0.3532	0.0398
B	-29.47	0.0315	0.708	0.674	0.655	0.641	0.607	0.581	0.4174	0.0426
C	-29.47	0.0315	0.748	0.723	0.708	0.698	0.673	0.656	0.4174	0.0426
D	-22.14	0.0292	0.708	0.674	0.655	0.641	0.607	0.581	0.5734	0.0329
E	-22.14	0.0292	0.708	0.674	0.655	0.641	0.607	0.641	0.4775	0.0330
F	-22.14	0.0292	0.708	0.674	0.655	0.641	0.641	0.641	0.6434	0.0252
G	-26.23	0.0278	0.632	0.588	0.561	0.543	0.496	0.461	0.4089	0.0332
H	-29.47	0.0315	0.748	0.723	0.708	0.698	0.673	0.656	0.4951	0.0279
I	-29.47	0.0315	0.708	0.674	0.655	0.641	0.607	0.581	0.6276	0.0283
J	-29.47	0.0315	0.708	0.674	0.655	0.641	0.607	0.581	0.4741	0.0342
K	-26.23	0.0278	0.689	0.658	0.639	0.626	0.595	0.572	0.4591	0.0279
L	-26.23	0.0278	0.759	0.733	0.717	0.706	0.677	0.654	0.6537	0.0267
M	-4.62	0.0098	0.759	0.733	0.717	0.717	0.677	0.654	0.6141	0.0257
N	-26.23	0.0278	0.689	0.658	0.639	0.626	0.595	0.572	0.4119	0.0295
O	-26.23	0.0278	0.689	0.658	0.639	0.626	0.595	0.572	0.3599	0.0312
P	-26.23	0.0278	0.619	0.577	0.552	0.535	0.492	0.459	0.3599	0.0312
Q	-4.62	0.0098	0.633	0.572	0.533	0.504	0.426	0.358	0.6537	0.0267
R	-4.62	0.0098	0.661	0.629	0.610	0.598	0.568	0.548	0.6141	0.0257
S	-4.62	0.0098	0.661	0.629	0.610	0.598	0.568	0.546	0.5218	0.0284
T	-4.62	0.0098	0.661	0.629	0.610	0.598	0.568	0.546	0.4119	0.0295
U	-4.62	0.0098	0.594	0.518	0.469	0.433	0.330	0.240	0,4119	0.0295

FLOW RATE AT VEADO STREAM AND TAIPAS CREEK SUB-BASINS

The discharge in North Area was measured *in situ* by the traditional technique that considers the variation of the of water level height in a certain position of the stream where the transversal section is known. The discharge in South Area was measured through a limnigraph settled in a local of known transversal section, and where continuous readings supplied the data of the variations of the water level that are necessary for evaluating the discharge.

However, despite these direct measurements, alternative approaches were utilized by Oliveira (2000) to estimate the discharge, which have been officially adopted in São Paulo State and are based on the technique suggested by Liazi et al. (1988). The method depends on the frequency of a certain discharge, i.e. the number of occurrences of that flow rate in a time interval and also of its reciprocal, i.e. the returning period (time interval between two occurrences of a certain discharge). Two types of flow rate may be estimated: the mean annual discharge, Q_m (in L/s), and the minimum annual discharge for consecutive 7 days over a returning period (T) of 10 years, $Q_{7,10}$ (in L/s).

According to Liazi et al. (1988), the steps to evaluate Q_m are:

(a) to find P, the average annual precipitation in the area (~1340 mm);
(b) to find the type of hydrographic basin according to the general classification for São Paulo State (in the present case, the Corumbataí river basin belongs to Tietê-Piracicaba hydrological region, which is represented by the letter "G");
(c) to estimate A, the basin area (in km^2);
(d) to find other parameters necessary for estimating Q_m, according to the equation Q_m (L/s) = (a + b × P) × A. In this study, *a* and *b* correspond to – 26.23 and 0.0278, respectively (Table QQ).

Thus, Q_m = 24.90 L/s in North Area, and Q_m = 337.5 L/s in South Area. This non-direct discharge evaluation in Taipas creek is reasonable if compared with values directly measured, as reported by Oliveira (2000). The flow rate of effluents released into Taipas creek is 8-9 L/s, that is significantly lower than the discharge of the drainage, as expected.

It is well known the fact that the rainfall can vary along each year, as well along several years, where the events may repeat regularly or randomly. In the case of hydrographic basins, it is important to calculate the probability of occurrence of periods of severe dry periods and the minimum expected discharge in these periods, as it may be critical the dilution of any pollutant in the basin. These factors may be calculated for returning periods of 10, 15, 20, 25, 50 or 100 years. Additionally, the minimum discharge is calculated for time intervals of 7, 30, 60 or 180 days. Obviously, these calculations depend on systematic reports performed for the basins in São Paulo State, however, this is a situation commonly not verified. According to Liazi et al. (1988), the equation for evaluating $Q_{7,10}$ is $Q_{7,10} = X_T \times Q_m (A + B) \times 0.80$, where X_T is the returning period T (for 10 years = 0.632) (Table 4), A and B are 0.4089 and 0.0332, respectively (Table 4), and 0.80 is the relation between the average minimum annual discharge for consecutive 7 days (Q_7) and the average minimum annual discharge for one month (Q_M). Thus, $Q_{7,10} = 5.56$ L/s in North Area, and $Q_{7,10} = 75.43$ L/s in South Area.

Figure 4. Location of the Guarani aquifer in South America.

Hydrogeological Aspects of the Areas Investigated

In terms of the hydrogeological context, the two areas investigated (North Area and South Area) are situated in the recharge zone of the huge Guarani aquifer of Triassic-Jurassic age that has continental dimensions, extending over some 950,000 km^2 within the Paraná sedimentary basin (Figure 4). The aquifer has an average thickness of 300-400 m, and is composed of silty and shaly sandstones of fluvial-lacustrine origin (the Pirambóia and Rosário do Sul Formations in Brazil and Buena Vista Formation in Uruguay), and variegated quartzitic sandstones accumulated by eolian processes under desert conditions (the Botucatu Formation in Brazil, Misiones Formation in Paraguay, and Tacuarembó Formation in Uruguay and Argentina) (Araújo et al., 1995). A thick basaltic package of the Serra Geral Formation (up to 1,500 m) overlies this aquifer, reducing its exposed areas in non-continuous elongated strips, 10-100 km wide, along the edges of the proper Paraná basin, where the longest strip stretches between the states of São Paulo and Santa Catarina, in Brazil. The aquifer overlies formations ranging from the igneous basement to the Paleozoic sediments of the Passa Dois and Tubarão Groups, being covered by Cretaceous sediments of the Bauru Group.

Data on water potentials have been obtained from exploration boreholes drilled for petroleum research and for exploiting the aquifer, whose depth range is 100-4600 m. The potentiometric surface of the water shows that ~ 70% of its total area has artesian conditions (Rebouças, 1988), and recharge occurs by direct infiltration of rainwater in the outcrop area, which is about 98,000 km^2 (Kimmelmann e Silva et al., 1989). The percolating water moves from the phreatic exposed areas that surround the entire basin towards its central part, where two outlets have been described: one allowing direct contact along the coast and the continental slope (a gap ~150 km wide in the crystalline framework in the states of Santa Catarina and Rio Grande do Sul in Brazil) and other masked by continental Quaternary deposits at SW end of the basin (NE Argentina). Some sporadic data indicate hydraulic conductivity values of 10^{-4}-10^{-5} m/s, effective porosity values of 10-20%, storage coefficient values of 10^{-3}-10^{-6} and average transmissivity of 10^{-3} m^2/s (Kimmelmann e Silva et al., 1989). The water is used extensively for drinking purposes, industrial activities and recreation (thermal swimming pools), with the yields of the wells varying from 10-150 m^3/h (phreatic exposed parts) up to 300-1000 m^3/h (confined artesian wells) (Gilboa et al., 1976).

MAJOR STEPS ON SAND PRODUCTION AT THE AREAS INVESTIGATED

Sand mining activities have been extensively held at the areas investigated (North Area and South Area). With the aim of producing sand for industrial use, the first step consists on the removal of vegetation from the mining area. Following this, the "sterile" material consisting on a well developed soil enriched in organic matter that covers the sandy layers is removed and stored for future use on relief remodeling at the mined area; the less developed soil exhibiting lower organic matter content is often utilized for civil construction after washing. Then, the sand layers are dug with appropriate vehicles, starting at weathered soil derived from Pirambóia Formation that is situated above the stone line and continuing up to the saprolite level under the stone line, as well up to the rock matrix level, i.e. Pirambóia Formation-Lower Unit. The recovered crude sand is conducted on trucks up to the treatment unit of the mining plant (Figure 5).

After arriving at the treatment unit, the crude sand is mixed with water at 1:1 proportion. The mixture is sieved for removal of roots, leafs, clay aggregates, quartz granules, and ferruginous/lateritic concretions. The next step is its insertion at the attrition cell where clay de-flakers like sodium hydroxide and sodium silicate are added. The clayey material (AFS clays) resulting from this first cleaning step is discarded into decantation pools. The selected material is transferred to a hydraulic classifier that utilizes a rising water flow for separating the sandy fraction according to three different sizes: gross, average, and fine.

Each sandy fraction is subjected to additional treatment processes through disaggregation cyclones that utilize the centrifuge force to cause the separation among fractions and the dewatering of the material. Additionally, some sand with specific grain size is submitted to a flotation process for removal of heavy minerals (for instance, magnetite, ilmenite, and hematite) that are undesirable for manufacturing several industrial products based on the use of sand. The flotation process basically consists on the injection of a great bubbles volume into mixture of sand and water. Organic compounds (in general, sulfonate salts) produce the bubbles, being carefully added in order to provide a chemical reaction when in contact with water. During their ascension to the liquid surface, the bubbles adhere to the mineral grains to be floated; then, they are separated from the liquid surface and discarded as waste at the decantation tanks.

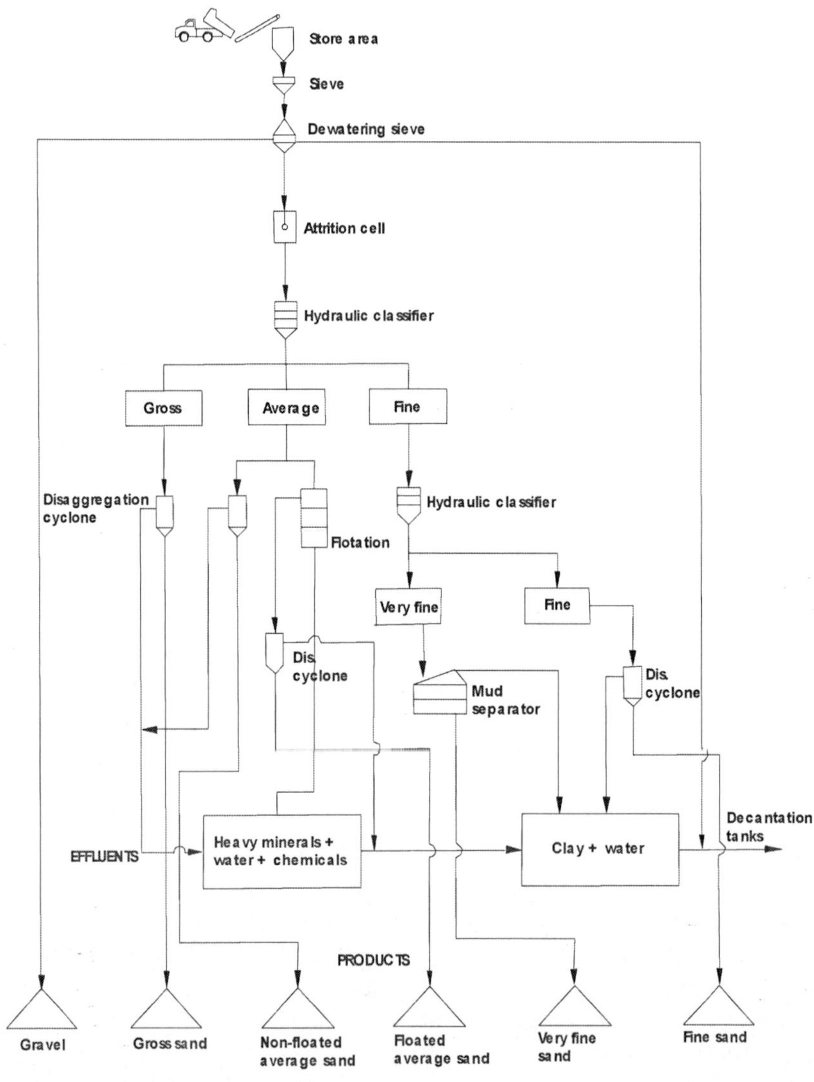

Figure 5. Simplified diagram of the treatment unit for sand utilized at South Area. Modified from Tandel (1993).

Thus, all residues generated during the sand mining activities are discarded into decantation pools and tanks. They mainly consist on sodium hydroxide, sodium silicate, organic and sulfonate salts, clayey minerals, AFS clays, and particulate material. In order to adjust the pH to 4.5-5.0, the effluents receive ferric chloride at the decantation pools, which accelerates the clay flocculation and sedimentation. The decantation tanks receive the water utilized on the sand washing that is enriched in AFS clays, as well de-flaked clays and chemicals utilized on the several involved steps. In the initial tanks occurs the sedimentation of the very gross and dense materials like fine sand and heavy minerals. In the following tanks occur the sedimentation of finer materials like silts and a portion of clays. Finally, in the last tanks, occurs the sedimentation of the clayey fraction, being generated a relatively limpid effluent that is re-pumped to the washing step (forming a closed circuit) or discarded into surface drainage. The sand mining plant at South Area (Figure 5) comprises 8 pools (average size = 2800 m^2) and 4 tanks (size between 3000 and 16000 m^2).

Major Tips for Evaluating the Hydraulic Conductivity

The hydraulic conductivity was evaluated by Oliveira (2000) in the unsaturated zone at both areas investigated (North Area and South Area) on using the following materials: five reservoirs of 200 liters each for providing water to the boreholes, one long wooden rod for scarification of the bore walls, one PVC tube (diameter = 12 cm) with a graduated scale and a transparent 0.5 cm diameter hose externally coupled for controlling the water level inside the tube, sealing mud (bentonite) between the PVC tube and bore walls, plastic hoses to conduct the water from the reservoirs up to the borehole and vehicle to transport the reservoirs up to the borehole.

Oliveira and Bonotto (2002) described the steps involved in the tests realized. After selecting the boreholes for determination of the hydraulic conductivity, they were cleaned on scarifying the sector of the walls to be tested. The purpose of such operation was to avoid the occurrence of plain surfaces resulting of the probe sampler rotation, which are less permeable than the real ones. Then, the depth of each borehole was measured, the PVC tube inserted in it up to 1 m above the ground level and the sealing mud added until the bottom of the tube.

The reservoirs filled with water were kept close to the top of the borehole, which was filled till the water level had reached the "zero" position in the lateral graduated scale, above the top of the tube. The water was supplied to the bore in order to assure a constant level at least during 20 to 30 minutes, period considered enough to saturate the moisture front around the bore. It is important reinforce that the saturation conditions of the massif must be carefully observed for checking if the water addition was much prolonged as the bore may cut holes dug by ants and mammals or a very fractured zone, under which situation it must abandoned and changed by another.

After reaching the saturation condition, the tube was filled slightly above the “zero” level. The water supplying was stopped and its level started lowering. The measurement of the time with a chronometer started when the water level reached the “zero” position. The readings corresponded to centimeters reached by the water level in the scale. For instance, after 5 s, it was performed a reading of the water level. Then, the bore was filled until the “zero” position and again the water supplying was stopped, determining how many centimeters the water level lowered, now in 10 s. Thus, the process was successively repeated increasing the time to 30 s, 1 min, 2 min, 5 min, 10 min, 15 min, 20 min, 30 min, etc. All data were recorded in an appropriate sheet for further evaluation of the hydraulic conductivity.

The right moment for finishing the measurements was decided on verifying the instant in which the time for some readings was long enough compared to the small lowering of the water level, at least during four successive measurements. Another way for checking this could be the insertion of the data obtained (time and water level variation) in a millimeter paper and the verification of the non-adjustment of the data to a log-curve (adjustment to a straight line). Under such condition, it would be additionally realized two or three measurements before finishing the essay, implying that the water flow became constant in the massif.

A linear regression involving the time and water level variation data allows finding the best straight line adjusted to the results, from which it may be evaluated the angular coefficient, a parameter necessary to apply the equation for determining the hydraulic conductivity (K). Some caution is needed to perform the regression because the initial points do not fit the straight line, even in mono-log graph, as they represent the phase in which the moisture front suffers influence of adsorption and capillary forces of the porous medium, so that the steady state of the hydraulic flow has not been reached. Consequently, the experienced hydrogeologist must disregard the points that certainly do not represent the actual behavior of the moisture front.

The same occurs when are generated non-typical points during the measurement, which are completely deviated from the general straight line, demonstrating that the moisture front has reached a point of high empties index. During some minutes, such condition raises the velocity vector of some specific restrict region of the front, consequently altering the indices values in a retroactive and momentary way.

The Hydraulic Conductivity Data

The hydraulic conductivity was evaluated as described by Oliveira (2000), according to the use of the following equation (ABGE, 1981):

$$K = dh/dt \times di^2 / 8\ H_o(DL)^{1/2}$$

where: K = hydraulic conductivity in the assayed sector (cm/s); h = depletion of the water level (NA) (cm); t = time for depletion of the water level (s or min); di = inner diameter of the PVC tube (cm); D = bore diameter (cm); L = distance between the bottom of the PVC tube and the bore bottom (cm); H_o = distance between the "zero" level and L/2 (cm).

Table 5. Parameters of the bores utilized in the essays conducted at the areas investigated. According to Oliveira and Bonotto (2002)

Parameter	Unit	North Area		South Area	
		Bore 1	Bore 2	Bore 1	Bore 2
PVC tube top – ground surface distance	m	0.71	0.70	0.73	0.74
PVC tube bottom – ground surface distance	m	0.72	0.73	0.70	0.69
PVC tube bottom – hole bottom distance	m	0.52	0.47	0.30	0.49
Assayed sector	m	0.72 -1.24	0.73-1.18	0.70-1.00	0.70-1.18
Bore diameter	cm	15	15	15	15
PVC tube - inner diameter	cm	12	12	12	12

Two essays were realized for each area investigated, as reported in Tables 5 and 6. The results obtained are illustrated in Figures 6-9, which indicate the following K values: 7.3×10^{-5} cm/s for bore 1 at North Area; 4.7×10^{-5} cm/s for bore 2 at North Area; 8.7×10^{-5} cm/s for bore 1 at South Area; 5.4×10^{-5} cm/s for bore 2 at South Area. According to the hydraulic conductivity classification for different materials (Castany, 1967), K values between 10^{-4} and 10^{-6} cm/s

(Table 7) would imply on a "bad" category that corresponds to the range of the values found at both areas investigated.

Table 6. Results obtained in the essays held at the areas investigated. According to Oliveira and Bonotto (2002)

Time	Cumulated time (min)	Water level variation (cm)			
		North Area		South Area	
		Bore 1	Bore 2	Bore 1	Bore 2
5 s	0.08	1.5	0.5	1.4	0.4
10 s	0.25	2.7	0.8	2.5	0.7
15 s	0.5	4.0	1.1	3.7	1.0
30 s	1	7.6	2.0	6.9	1.9
1 min	2	13.5	4.5	13.1	3.8
2 min	4	28.6	8.0	26.5	7.3
3 min	7	40.3	10.9	36.7	9.9
5 min	12	57.9	16.5	53.6	15.0
10 min	22	88.5	28.5	81.0	26.0
15 min	37	109.0	38.0	99.5	34.0
20 min	57	114.5	44.5	107.5	39.5
25 min	82	121.5	55.5	111.8	47.9
30 min	112	127.0	57.5	116.4	51.3

Tablc 7. Classification of different materials according to the hydraulic conductivity values by Castany (1967)

Hydraulic Conductivity (cm/s)											
10^2	10^1	10	10^{-1}	10^{-2}	10^{-3}	10^{-4}	10^{-5}	10^{-6}	10^{-7}	10^{-8}	10^{-9}
Gravel			Pure sand			Very fine sand			Pure clay		
Coarse sand			Sand and gravel			Silt and mixing of					
without fines			without fines			sand and clays					
very good			good			bad			impermeable		

The coefficients obtained at both areas indicate that the hydraulic flow is higher in the upper than in the lower portion of the soil profiles. This is a consequence of the major features of the pedogenetic profile that originated more laterized soils with more porous structures in the upper portions of the profile and more structured, clayey and compact soils in its lower levels. In this case, tectonics forces contributed to locate similarly evolved soils in different levels, and, consequently, it is not appropriate extrapolate hydraulic conductivity values from one horizon to another, even when situated at the same topographic level and possessing equivalent lithology, as there are faults of several meters that may result in differences on the hydraulic behavior.

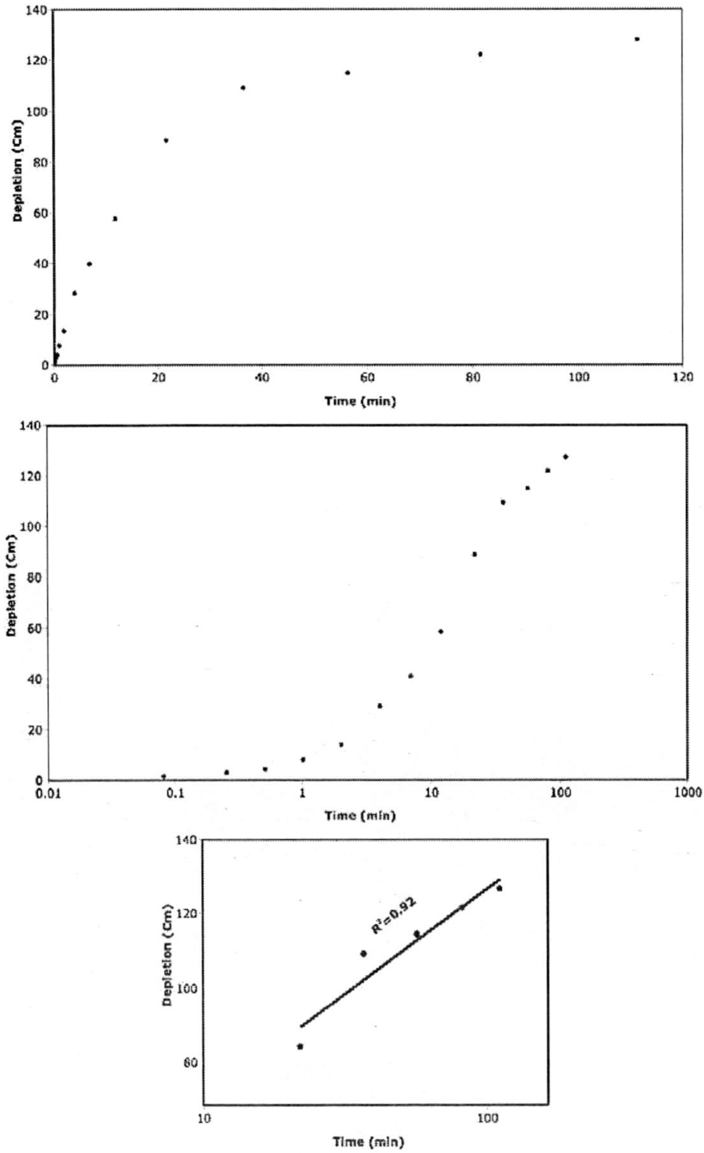

Figure 6. Results obtained in the essays conducted at bore 1 – North Area.

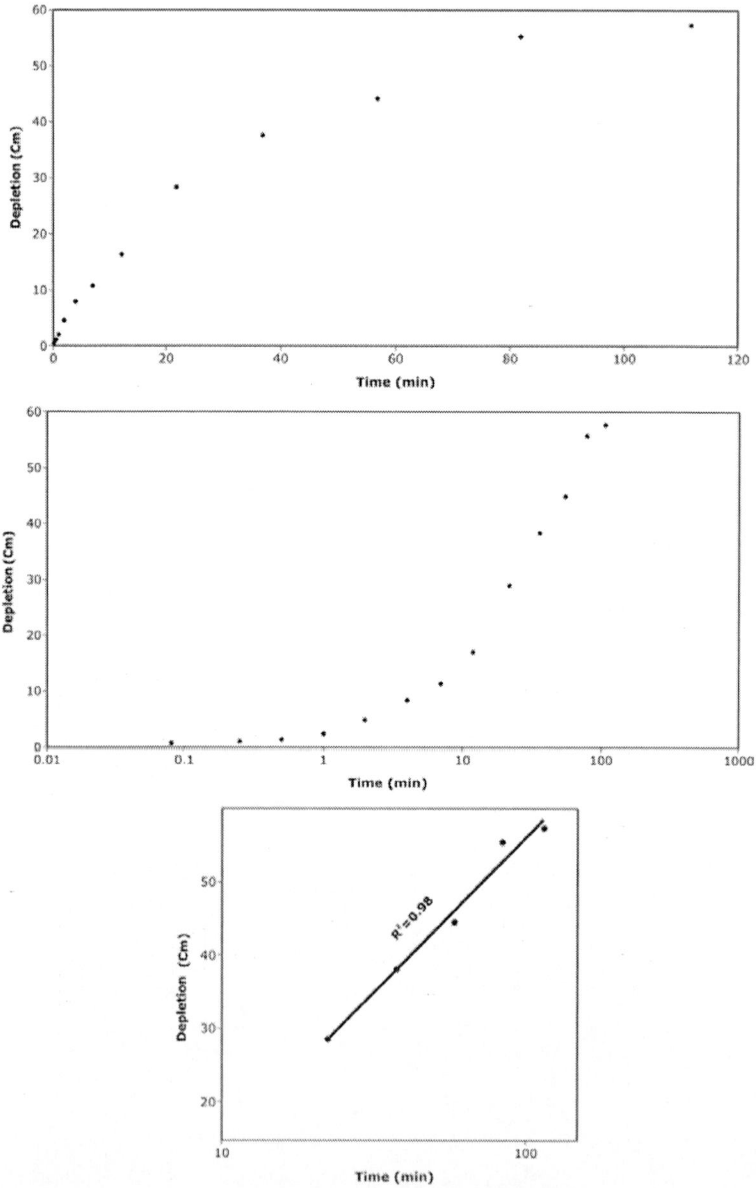

Figure 7. Results obtained in the essays conducted at bore 2 – North Area.

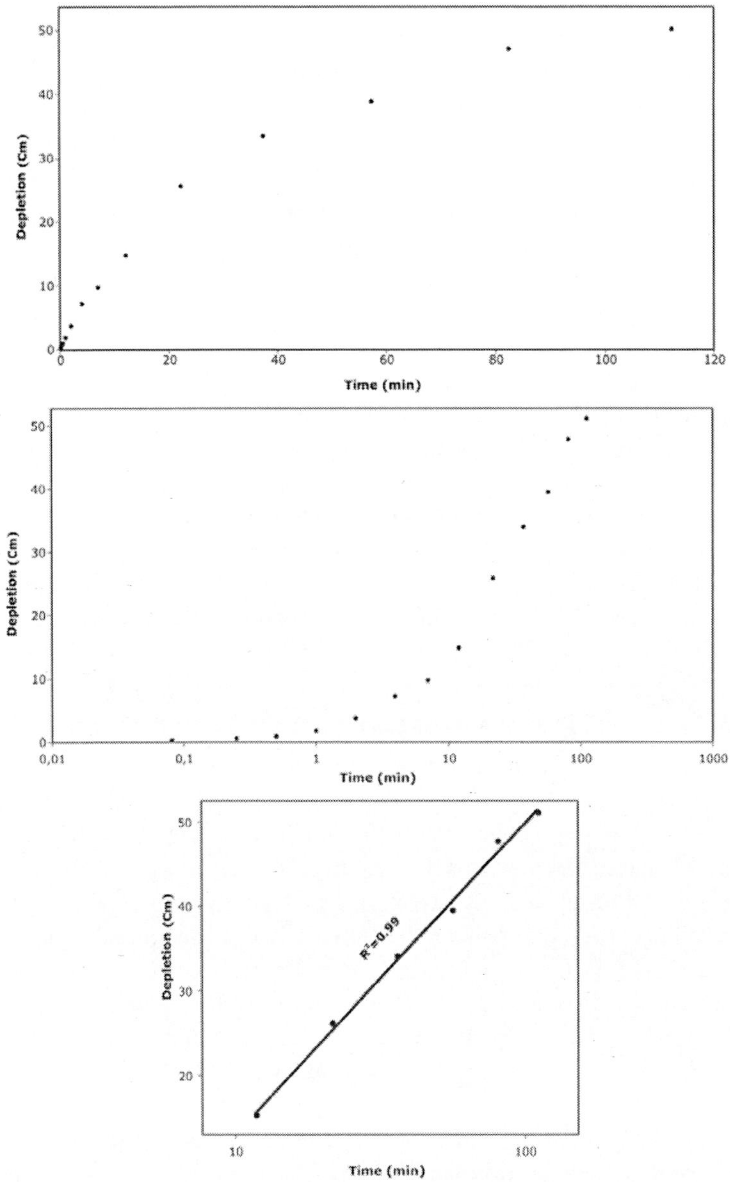

Figure 8. Results obtained in the essays conducted at bore 1 – South Area.

Another consequence of the difference on the hydrodinamic behavior of the underground flow is that it is unconfined and less subjected to lithological and pedological controls in the case of North Area compared to South Area. This may be clearly verified by the groundwater flow lines that are more homogeneous in North Area, exhibiting lower hydraulic gradient and whose horizontal vector tends to be directed towards the local baseline level that is Veado Stream. Such deflection assures that there are no lithological and structural "barriers" responsible to reduce the hydraulic flow.

However, the flowlines in South Area are less homogeneous, have higher gradient and suffer several inflexions in the direction vectors, where some are SE-oriented, in opposition to the general direction. This occurs due to the lithological and pedological controls, as well by the tectonic and structural ones, because there is a great faulting in the area represented by the channel of the Grand Canyon Current, which extends in the NW direction and is not visible neither by surface geological mapping nor by aerial photos as it is covered by creep soils and colluvium existing in the local. On the other hand, it was identified by the different behavior of groundwater flowlines. These observations are reinforced by the fact that there is another faulting in the area exhibiting the same behavior, which is located about 1 km towards NE.

Hydrochemistry and Water Quality at North Area

The sand mining activities at both areas investigated potentially can be a source of anthropogenic impacts, since several chemicals are used for the treatment of natural sand like HCl, H_2SO_4, NaOH and Na_2SiO_4, generating mine tailings that can pollute the surface and underground hydrological resources. The values obtained for various parameters in waters from the North Area were compared with those defined by the national standards for drinking water. The study of the hydrologic environment around the waste disposal area involved the sampling of groundwaters from 3 boreholes (Table 8) drilled for characterizing the subsurface flow and of surface waters from Ponte Funda Stream that receives most of the area drainage (Figure 10). The aquifer system developed in the weathered cover over Pirambóia Formation, being performed three campaigns for water sampling.

The samples were stored in polyethylene bottles, and depending on the requirements of the analysis, they were distributed as unfiltered and unpreserved (i.e. for temperature, pH, Eh, dissolved O_2, conductivity and bicarbonate determinations), filtered through 0.45 μm membrane and unpreserved (i.e. for major ions analyses) and filtered and preserved with different acids (i.e. for trace element determinations). Standard analytical techniques were used for obtaining the composition of the major and minor elements in waters, for example, methyl orange end-point titration, potentiometry, flame photometry, ion selective electrode and inductively-coupled plasma emission. The results of the measurements are reported in Table 9.

The studied groundwaters belong to a low-temperature aquifer system, whose Eh-pH field indicates it is transitional, tending to be reducing in character. When the chemical data obtained for the surface and groundwaters are compared with the maximum (minimum in the case of dissolved oxygen) permissible concentration limits in drinking water established by the national standards NTA-60 (São Paulo State Register 12486 published on 20 October 1978) and CONAMA No. 20 (National Register for fresh waters belonging to class 2, published on 18 June 1986), it is possible to verify that the analytical techniques utilized for the measurements of NH_4, P, Cd and Pb have detection limits above the permissible levels. With the exception of pH, all other values obtained for the investigated parameters don't suggest the occurrence of any environmental impact related to the sand mining activities.

The pH values could be attributed to some anthropogenic input, however, the cation Na^+ and anions SO_4^{2-} and Cl^-, which are potentially important to affect the area, are present in very low concentrations, confirmed by the low values of conductivity and dry residue, reflecting, in the case of the groundwaters, the composition of rainwater with small amounts of silica added from contact with the aquifer.

Table 8. Characteristics of the boreholes (BH) studied at North Area

Parameter	Unit	BH 1	BH 2	BH 3
Diameter	mm	75	75	75
Altitude of the top	m	711.5	688.0	685.0
Total length	m	16.0	5.5	5.5
Water table depth	m	13.20	2.96	1.60

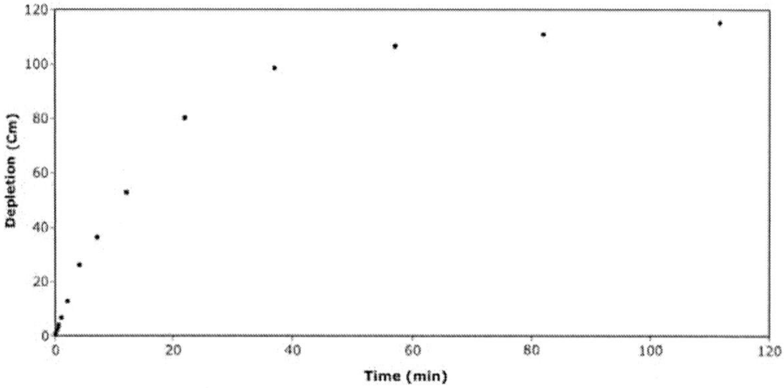

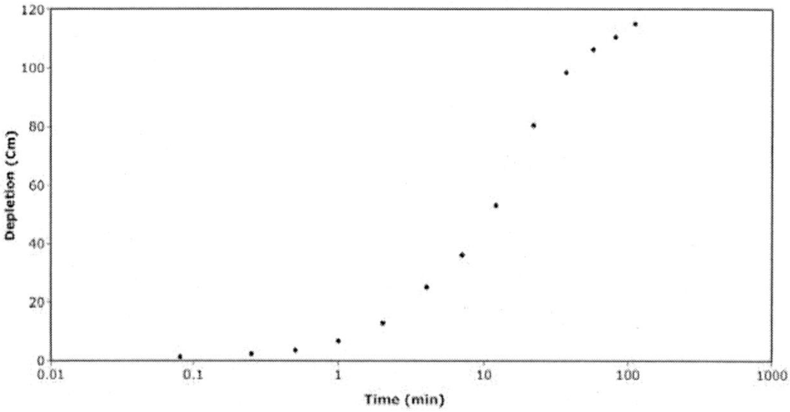

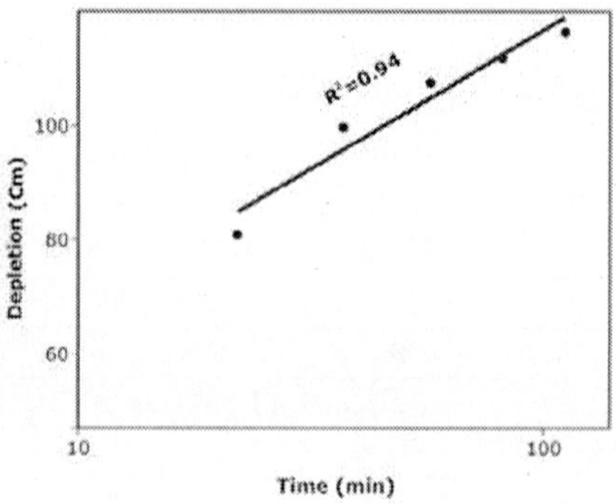

Figure 9. Results obtained in the essays conducted at bore 2 – South Area.

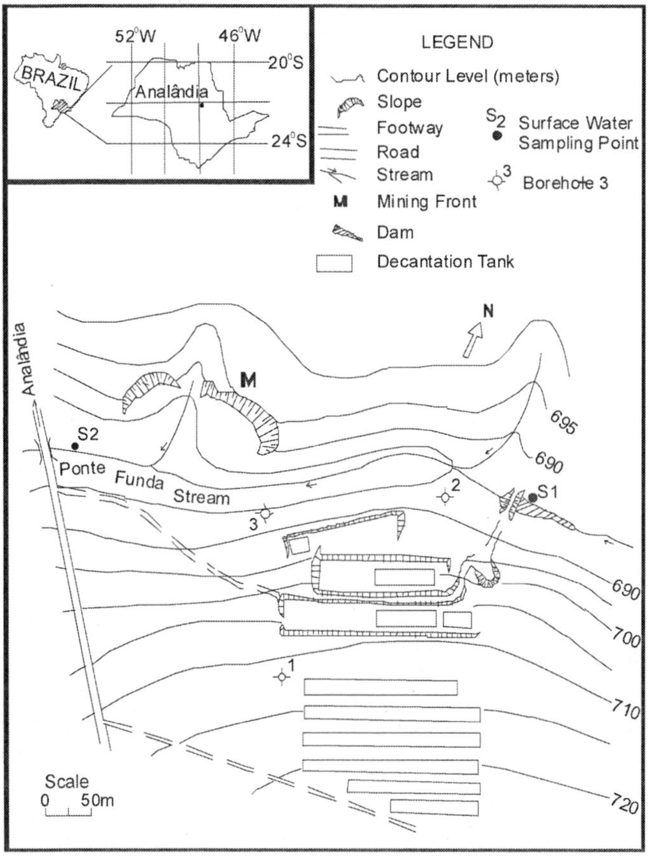

Figure 10. Location of the mine tailings area at North Area in Analândia city, São Paulo State, Brazil, and of the sampling points for ground and surface waters.

Such pattern of chemical data is not much amenable to plot on a standard Piper (1944) diagram, because there is a lack of the preponderance of typical anions and cations, and the mixed character is identified in some situations (Figure 11). The rainwater interaction with the weathered mantle developed over Pirambóia Formation when it enters the aquifer by percolating through the unsaturated zone until it meets the water table constitutes another explanation for the low pH values, because the pH of several soil suspensions representative of horizons from three pedological profiles was determined both in water and in KCl solution and the most of the obtained values are strongly acid (pH<4.5) or moderately acid (pH between 4.5 and 5.5).

Table 9. Chemical analyses of groundwaters (BH) and surface waters (S) at the North Area in Analândia city, São Paulo State, Brazil

Parameter	Unit	BH 1	BH 2	BH 3	S1	S2
Temperature	°C	25	26	26	31	26
Dissolved O_2	mg/L	7.5	7.1	4.5	10.0	8.0
pH	-	4.2 5.2	5.0	3.9 4.6	4.4	4.8
Eh	mV	+183		+217		
Conductivity (25°C)	μS/cm	50	50	40	40	40
Silicon	mg/L	3.0	4.2	3.3	3.7	3.3
Bicarbonate	mg/L	2.9	5.1	1.0	8.0	12.0
Sulfate	mg/L	<2.0	<2.0	<2.0	<2.0	<2.0
Chloride	mg/L	1.2	0.9	0.3	0.4	0.6
Fluoride	mg/L	0.06	0.05	0.04	0.05	0.04
Nitrate + Nitrite	mg/L	0.03	0.09	0.02	0.03	0.03
Ammonium	mg/L	<0.2	<0.2	<0.2	<0.2	<0.2
Phosphorus	mg/L	<0.1	<0.1	<0.1	<0.1	<0.1
Boron	mg/L	<0.02	<0.02	<0.02	<0.02	<0.02
Aluminum	mg/L	<0.1	<0.1	<0.1	<0.1	<0.1
Sodium	mg/L	0.8	1.6	0.3	0.1	0.1
Potassium	mg/L	0.5	0.6	0.2	0.3	0.2
Calcium	mg/L	1.6	1.5	0.4	0.1	0.1
Magnesium	mg/L	0.1	0.1	<0.1	<0.1	<0.1
Barium	mg/L	0.02	0.03	0.02	<0.02	<0.02
Total iron	mg/L	<0.02	<0.02	<0.02	<0.02	<0.02
Manganese	mg/L	<0.02	0.04	<0.02	<0.02	<0.02
Zinc	mg/L	0.02	<0.02	0.02	<0.02	<0.02
Copper	mg/L	0.02	0.02	0.02	<0.02	<0.02
Lead	mg/L	<0.1	<0.1	<0.1	<0.1	<0.1
Cadmium	mg/L	<0.02	<0.02	<0.02	<0.02	<0.02
Total chromium	mg/L	<0.05	<0.05	<0.05	<0.05	<0.05
Nickel	mg/L	<0.02	<0.02	<0.02	<0.02	<0.02
Dry Residue(105°C)	mg/L	40	36	32	24	20

If the chemical data obtained for the waters are used to study stability relations of mineral phases in the systems K_2O-Al_2O_3-SiO_2-H_2O, CaO-Al_2O_3-SiO_2-H_2O and Na_2O-Al_2O_3-SiO_2-H_2O (Garrels and Christ, 1965), it is observed that they fall into the kaolinite field of stability (Figure 12), information confirmed by the use of X-ray diffraction in the mineralogical identification of samples from the weathered cover developed over Pirambóia Formation, since kaolinite was identified as the second most abundant mineral phase (SiO_2 was the first) (Jiménez-Rueda, 1994).

The total Fe concentration reported in Table 9 for BH3 doesn't reflect the actual value, because after exposure to the atmosphere, an originally light and colorless high-Fe water turned red as ferric species formed and a reasonably high amount of material precipitated, implying that the measurement only refers to the supernatant phase after filtering in 0.45μm membrane. The major Fe-bearing mineral in the area is limonite, which has been commonly identified in the weathered mantle developed over Pirambóia Formation (Tandel, 1993). When a solution containing Fe^{2+} is neutralized (pH increases), or aerated (Eh increases), or both, amorphous hydrous Fe oxides precipitate, reordering, subsequently, into more crystalline Fe oxides such as goethite or hematite (Faure, 1991), pattern not confirmed by the reported data for waters of BH1 and BH3 since, if the direction of the flow-through water from BH1 up to BH3 is taken into account, then, it is possible to verify that the pH decreases, Eh increases, and aeration (dissolved O_2) decreases. The lower aeration related to the Eh rise in BH3 probably can be attributed to the dissolved O_2 consumption for the oxidation of Fe^{2+} to Fe^{3+}, corresponding to 34 mV the difference on Eh values measured in BH1 and BH3, which agrees very reasonably with 32 mV evaluated from the theoretical relationship between the Eh, pH and fugacity of O_2 (Faure, 1991), i.e.:

$$Eh = 1.23 + 0.01479 \log [O_2] - 0.05916 \, pH$$

Chapter 4

U-HYDROGEOCHEMISTRY IN GUARANI AQUIFER

The natural radioelement uranium is widely distributed in crustal rocks, and under conditions present at the earth's surface, it tends to be a mobile element. ^{238}U is the principal isotope of natural U (99.3% abundance) and is the parent nuclide in the mass number (4n+2) decay series, the longest known series. The isotope ^{234}U is radiogenic and the decay chain from ^{238}U through ^{234}U proceeds as follows: ^{238}U (4.46 Ga, α) $\rightarrow$ ^{234}Th (24.1 days, β^-) $\rightarrow$ ^{234}Pa (1.18 min, α) $\rightarrow$ ^{234}U (248 ka, α) $\rightarrow$ ....

The isotopes ^{238}U and ^{234}U are in secular equilibrium in all minerals and rocks greater than one million years old and that are closed systems for U. The $^{234}U/^{238}U$ activity ratio (AR) is therefore unity in the bulk of such systems, however, rock-water interaction frequently results in ARs for dissolved uranium which are greater than unity (Osmond and Cowart, 1976; Ivanovich and Harmon, 1982). Alternative mechanisms which have been suggested for the generation of such enhanced ARs include preferential chemical dissolution of ^{234}U (Rosholt et al., 1963) and alpha-recoil release of ^{234}Th at the rock-water interface (Kigoshi, 1971).

Uranium in the hydrologic environment is of special interest because of its economic importance and its chemical and radiotoxicity and that of some of its daughter nuclides. Worldwide soluble concentrations generally range from 0.1 to 10 μgL^{-1} in rivers, lakes and groundwaters (Fritz and Fontes, 1980; Ivanovich and Harmon, 1982). In order to properly evaluate the dissolution of U it is important to take into account several parameters, among them, the temperature, redox conditions and pressure that can affect significantly the kinetics and thermodynamic properties of the mineral phases, with implications for their dissolution or precipitation.

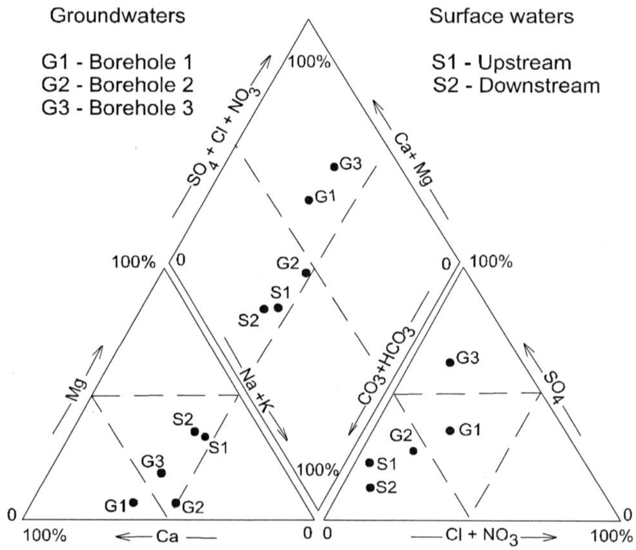

Figure 11. Chemical data for the studied waters plotted on a Piper (1944) diagram.

Table 10. Summary of hydrochemical data obtained for groundwater samples from Guarani aquifer

Parameter	Unit	Range	Parameter	Unit	Range
Temperature	°C	18 to 70	Sodium	mg/l	0.3 to 322
pH	-	4.0 to 9.9	Potassium	mg/l	0.04 to 5.3
Eh	mV	-97 to 630	Calcium	mg/l	0.2 to 45
Dissolved O_2	mg/l	0.8 to 9.5	Magnesium	mg/l	<0.1 to 13
Conductivity	µS/cm	6 to 1480	Silicon	mg/l	4.5 to 28.2
Dry residue	mg/l	100 to 1000	Boron	mg/l	<0.005 to 1.8
Bicarbonate	mg/l	0.1 to 318	Strontium	mg/l	<0.01 to 1.7
Carbonate	mg/l	6 to 145	Barium	µg/l	1 to 407
Chloride	mg/l	<0.15 to110	Zinc	µg/l	<4 to 91
Sulfate	mg/l	<0.3 to351	Copper	µg/l	<2 to 31
Nitrate	mg/l	0.5 to 10	Uranium	µg/l	0.004 to 15.4
Fluoride	mg/l	<0.02 to 8.8	$^{234}U/^{238}U$	A.R.	1.0 to 27.9

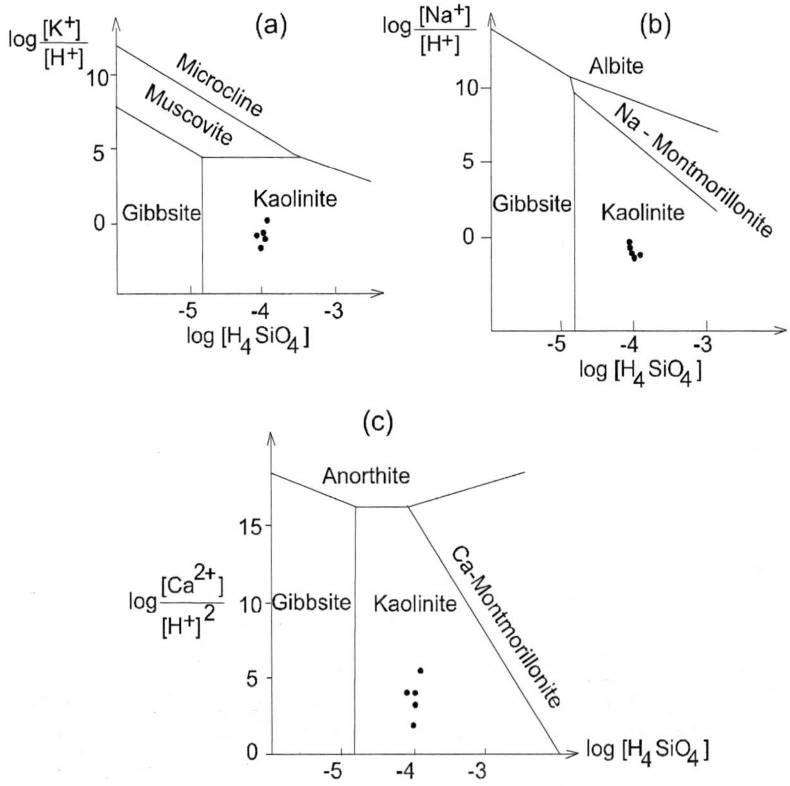

Figure 12. Chemical data for the studied waters plotted on diagrams representing the stability relations of some mineral phases in the systems a) K_2O-Al_2O_3-SiO_2-H_2O, b) Na_2O-Al_2O_3-SiO_2-H_2O, c) CaO-Al_2O_3-SiO_2-H_2O at 25°C and 1 atmosphere (Garrels and Christ, 1965).

The U-hydrogeochemical behavior in Guarani aquifer was evaluated on conducting a sampling campaign held in 66 localities situated at São Paulo, Mato Grosso do Sul, Paraná, Santa Catarina and Rio Grande do Sul States in Brazil, where 77 groundwater samples were collected from free-flowing and pumped tubular wells for physical and chemical analyses, including uranium. The available data describing the wells allowed to estimate the lithostatic pressure, P, from the equation (Castany, 1982): $P = Pa + \rho gh$, where Pa is the atmospheric pressure, ρ is the average density of the terrain, g is the gravity acceleration, and h is the depth of the top of the aquifer.

The groundwater samples (19-20 kg) were stored in polyethylene bottles, one unfiltered and unpreserved aliquot was used for temperature, dissolved oxygen, pH, Eh redox potential, conductivity, bicarbonate, and carbonate determinations, whereas other aliquot was filtered through 0.45 μm membrane and preserved with HCl or HNO_3 for the evaluation of uranium isotopes and other elements and compounds. Standard analytical techniques were used for obtaining the composition of the major cations and anions in waters, for example, methyl orange end-point titration, potentiometry, ion selective electrode, and spectrophotometry.

The aliquots for U analysis were acidified to pH less than 2 on using HCl, about 500 mg of $FeCl_3$ plus 3.39 dpm of ^{232}U were added, and U was co-precipitated on $Fe(OH)_3$ by increasing the pH to 7-8 through addition of concentrated NH_4OH solution; the precipitated was recovered, dissolved in 8M HCl and Fe^{3+} was extracted into an equal volume of isopropyl ether. The acid solution U-bearing was purified by anion exchange, first on a Cl^- and then on a NO_3^- column of Rexyn 201 50-100 mesh resin. U was finally eluted from the NO_3^- column with 0.1 M HCl and after evaporation to dryness was dissolved in 10 ml of 2M $(NH_4)_2SO_4$ electrolyte and transferred to an electrodeposition cell. The pH was adjusted to 2.4 and electrodeposition of U on a stainless steel planchet was complete after 3 hours at a current density of 1 Acm^{-2}.

The U content was measured by alpha spectrometry. The α-activities were determined with two 0.1 mm depletion depth, 200/450 mm^2 area silicon surface barrier detectors. The spectra for natural U and ^{232}U tracer extracted were recorded on a EG&G ORTEC 919 Spectrum Master Multichannel Buffer. The Decision Level L_c (Currie, 1968) for acceptance of a positive measurement in the ^{238}U and ^{232}U energy regions was 0.00082 and 0.00225 cpm, respectively. The concentration data were calculated by isotope dilution from the counting rates of ^{238}U and ^{232}U peaks, where the analytical details for these measurements were reported elsewhere (Ivanovich and Harmon, 1982).

The results obtained in the analyses are summarized in Table 10. The evaluation of the CO_2 partial pressure of the samples was performed from the data for bicarbonates and carbonates on using the WATEQ 4F aqueous speciation geochemical modeling software (Truesdell and Jones, 1974), package that also allowed to calculate the ionic strength of the samples, and the mineral saturation index (SI) for calcite, dolomite, and amorphous silica, as reported in Table 11.

Table 11. Ionic strength, CO_2 partial pressure, SI calcite, SI dolomite and SI amorphous SiO_2 in groundwaters from Guarani aquifer

Sample No.	Ionic strength (×10-3)	CO2 partial pressure (×10-3 atm)	SI calcite	SI dolomite	SI am. SiO2
01	5.37	0.0376	0.370	0.159	-0.740
02	-	-	-	-	-
03	8.46	0.342	-0.872	-1.836	-0.873
04	4.72	0.0195	0.156	-0.060	0.288
05	3.14	0.0199	-0.023	-0.454	-0.621
06	9.02	0.8410	0.630	1.398	-
07	10.99	0.4990	0.532	-0.194	-0.772
08	5.16	0.0660	0.143	0.262	-0.764
09	5.37	0.0763	0.251	0.131	-0.749
10	3.31	0.1490	-0.263	-0.894	-0.567
11	0.20	93.4000	-5.898	-11.831	-0.629
12	0.21	198.000	-7.467	-14.375	-0.897
13	5.12	0.0901	0.236	0.042	-1.469
14	6.46	0.0393	0.635	1.052	-0.988
15	0.49	69.8000	-4.618	-9.184	-0.205
16	0.56	63.1000	-3.775	-8.011	-0.234
17	7.89	0.2350	0.584	0.237	-0.739
18	4.33	0.4650	-0.257	-0.507	-1.183
19	4.17	0.1770	-0.303	-0.761	-1.514
20	4.75	0.1310	0.037	0.226	-0.748
21	4.85	0.1440	0.325	0.203	-0.624
22	0.59	0.0046	-1.879	-4.374	-0.488
23	5.03	0.1580	0.404	0.821	-0.561
24	2.40	34.3000	-1.616	-3.481	-0.088
25	5.08	0.5390	0.039	-0.108	-0.686
26	7.23	0.0250	0.756	1.419	-0.858
27	8.66	0.1450	-0.204	-0.098	-
28	1.44	0.1970	-0.109	-1.353	-0.597
29	1.44	0.1190	-0.364	-1.100	-0.471
30	3.11	16.2000	-0.890	-2.726	-0.434
31	1.16	-	-	-	-0.235
32	0.25	23.7000	-5.237	-10.375	-0.504
33	0.27	9.0500	-4.401	-8.604	-0.742
34	0.60	159.0000	-4.679	-9.385	-0.741
35	0.58	145.0000	-4.018	-8.624	-0.402
36	0.89	73.6000	-3.476	-7.270	-
37	0.81	6.2300	-2.423	-6.046	-1.746

Table 11. (Continued).

Sample	Ionic	CO2 partial	SI	SI	SI
38	0.34	64.2000	-5.654	-10.158	-0.381
39	0.37	36.7000	-4.876	-9.350	-0.441
40	0.44	55.6000	-5.381	-9.555	-0.395
41	0.47	0.6750	-3.864	-6.633	-0.462
42	2.47	0.2880	0.371	-0.282	-0.618
43	0.70	1.9200	-2.951	-4.715	-0.726
44	2.38	3.4600	-0.442	-1.196	-0.860
45	3.88	21.6000	-0.907	-1.994	-0.843
46	0.44	12.8000	-3.382	-7.069	-0.559
46	1.20	21.9000	-2.134	-4.962	-0.427
48	0.42	30.2000	-3.862	-7.838	-0.520
49	0.69	7.3900	-2.785	-5.756	-0.342
50	0.37	24.3000	-4.481	-8.644	-0.324
51	3.14	66.7000	-1.504	-4.057	-
52	4.39	20.0000	-0.591	-1.909	-0.128
53	3.45	35.7000	-1.151	-2.860	-0.356
54	2.60	72.0000	-1.990	-4.343	0.025
55	0.54	0.1970	-2.214	-3.780	-0.378
56	2.76	27.0000	-1.297	-3.303	-0.225
57	1.11	16.8000	-2.371	-5.482	-0.148
58	1.90	2.0600	-0.567	-2.544	-0.172
59	0.32	10.9000	-4.712	-9.012	-0.559
60	1.82	79.9000	-2.428	-5.088	-0.048
61	2.42	14.0000	-1.199	-2.979	-0.733
62	1.89	20.8000	-1.624	-3.973	-0.460
63	0.45	0.8570	-3.071	-5.291	-
64	0.66	313.0000	-5.653	-11.254	-
65	0.60	0.7920	-2.725	-4.413	-
66	3.50	0.0803	-0.013	-0.829	-
67	0.65	12.2000	-4.356	-7.414	-
68	1.89	1.9800	-0.420	-2.900	-
69	15.80	37.0000	-1.191	-3.748	-0.183
70	0.68	2.5400	-3.306	-5.538	-
71	6.29	6.45	0.064	-0.127	0.023
72	8.45	0.6240	1.254	2.175	-
73	0.34	1.9500	-3.214	-6.528	-
74	0.60	0.5800	-2.358	-3.529	-
75	1.99	0.6040	-0.159	-0.835	-0.312
76	1.33	5.9600	-1.705	-3.788	-0.138
77	0.66	14.1000	-3.907	-7.246	-

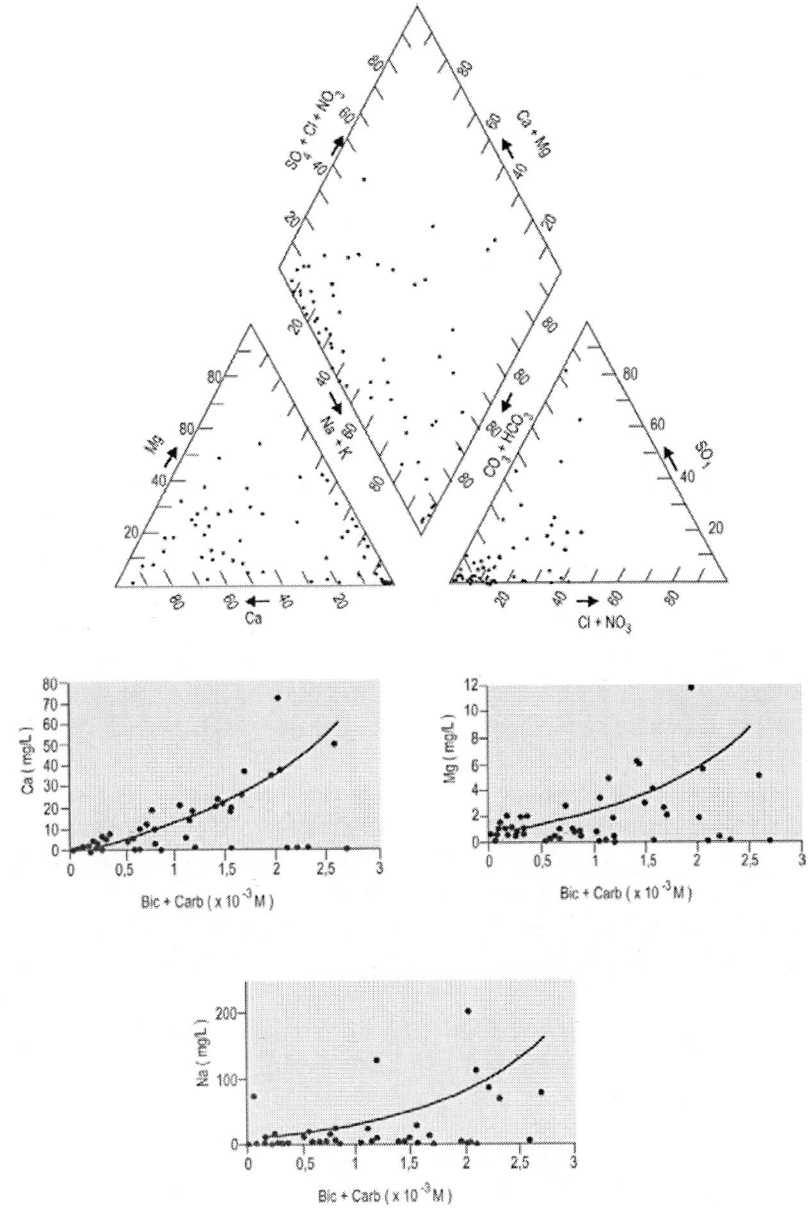

Figure 13. Chemical data for groundwaters from the Guarani aquifer having calculated negative SI-calcite values plotted on a Piper (1944) diagram and the amount of Ca, Mg, and Na dissolved in relation to bicarbonate + carbonate dissolution.

The occurrence of water at elevated temperatures (above 40°C) was verified due to the great depths the aquifer reaches (almost 2 km) and the thick confining basaltic cover. A linear relationship (r = 0.94) was observed between the temperature and lithostatic pressure, confirming the results obtained by others investigators (Silva, 1983; Teissedre and Barner, 1981) and relating the groundwater flow from the border of the basin towards its central part, in the direction of the dip of the geological units, according to the natural geothermal gradient of about 1°C/35 m.

The dissolution of silica and carbonates in the Guarani aquifer can be characterized by negative SI values, whereas their precipitation may be indicated by positive SI values. Because the precipitation of carbonates is a very important phenomenon occurring at many wells drilled in the Guarani aquifer, it is convenient to discuss the hydrochemistry taking into account this process. Thus, the SI-calcite values reported in Table 11 will be utilized for such purpose.

The chemical data obtained for samples characterized by negative SI-calcite values were initially plotted on a standard Piper (1944) diagram, being possible to identify the mixed character in terms of dissolved anions and cations, as well two principal hydrochemical facies: Ca^{2+}-HCO_3^- and Na^+-HCO_3^- types (Figure 13). Thus, the major cations Na, Ca, and Mg are dissolved by first-order chemical leach, where the leach rates greatly increase with the (bi)carbonate content in solution (Figure 14). The main mineral phases responsible for the release of these ions into solution are probably (hydro)magnesite, dolomite, calcite, aragonite, nahcolite, trona, natron, thermonatrite, huntite, nesquehonite, and artinite.

The chemical data obtained for samples characterized by positive SI-calcite values were also plotted on a standard Piper (1944) diagram, being possible to verify in this case that the waters are mixed and HCO_3^--dominated in terms of dissolved anions, whereas they are greatly Na^+-dominated in terms of dissolved cations (Figure 14). Some negative SI-dolomite values were calculated for samples having positive SI-calcite values (Table 11), suggesting that the dissolution of dolomite tend to modify the dominance of Na in the liquid phase, like observed for three samples (Figure 14). Thus, the data separation according to their positive SI-calcite values indicates preponderantly the occurrence of precipitation of Ca and Mg with (bi)carbonate ions, however, despite this, the same is not observed for sodium that is being dissolved by first-order chemical leach, with the leach rate increasing with the content of (bi)carbonate in solution (Figure 14).

Some Effects of the Temperature on the Hydrochemistry

The properties of silica and carbonate phases, as well the temperature, pH, redox potential (Eh), and CO_2 partial pressure are useful parameters for understanding the hydrogeochemical behaviour of dissolved uranium isotopes. Depending of the temperature of the water, kinetics and properties of the carbonate phase considered, then, dissolution or precipitation of the mineral may be occurring. For instance, when calcite is taken into account, a saturated solution of this mineral in equilibrium with CO_2 at 3×10^{-3} atm contains about 75 mg/L of Ca^{2+} at 5°C, but only 40 mg/L at 30°C (Faure, 1991), showing, as a result, that dissolution of this mineral is favored at lower temperatures, whereas precipitation is expected at higher temperatures. In spite the solubility of calcite decreases with increasing temperature, this, in fact, is not the rule for all carbonates, since some of them exhibit an opposite behavior, justifying the data pattern verified in Figure 14.

Because the dissolution/precipitation of carbonates is dependent of the temperature, the data obtained were initially represented against this parameter for negative SI-calcite values (Figure 15). As a consequence of the relationship between temperature and geostatic pressure, a similar trend to that reported by Bonotto (1999a) was verified, indicating that the increase in temperature causes a decrease in the CO_2 partial pressure of the analyzed thermal and non-thermal groundwaters (Figure 15a). The same trend would be obtained if the dissolved oxygen data were plotted against temperature instead geostatic pressure like proposed by Bonotto (1999b).

The reducing conditions (Figure 15b) and basicity (Figure 15c) also rise in accordance with the increase of temperature (up to 70°C), accompanying the groundwater flow from the border of the basin towards its central part, in the direction of the dip of the geological units. The dissolved Ca and Mg contents decreased with the temperature (Figs. 15d and 15e), suggesting that the dissolution of carbonates is decreasing in accordance with processes characterized by parabolic kinetics, in which the rock matrices of the aquifer are dissolved by first-order chemical leach (decreasing leach rates with time).

The supersaturated samples characterized by positive SI-calcite values also exhibited similar decrease of the redox potential (Figure 16b), dissolved Ca content (Figure 16d), and dissolved Mg content (Figure 16e) with the increasing temperature, however, the CO_2 partial pressure (Figure 16a) and pH (Figure 16cd) of these samples were not significantly affected by temperature.

URANIUM ISOTOPES AND AGES

The evaluation of the exact age of groundwater masses is difficult not only because analytical techniques may yield data which are difficult to relate to the geochemical changes during ground water evolution but also because it must be generally assumed that the samples collected are representative of a single ground water body. Conventional ^{14}C dating has been utilized for ground waters of up to 30,000 to 40,000 years old (Fritz and Fontes, 1980; Ivanovich and Harmon, 1982), however, in order to date older hydrological systems, other tools have been used, for instance, the isotopic compositions of dissolved uranium (Ivanovich and Harmon, 1982). Uranium in solution may be depleted or enriched, depending on oxidation-reduction conditions (Osmond and Cowart, 1976; Langmuir, 1978; Bowie and Plant, 1983). It is an element very sensitive to modifications on redox potentials (Eh), since if oxidizing conditions prevail, active solution of U occurs, and if the redox character changes towards reducing conditions (lower values of Eh), then, deposition of U takes place.

Thus, in order to use the U-isotopes chronological modeling it is necessary to obtain higher U contents and lower ARs values when oxidizing conditions prevail in the recharge zone of an aquifer, as well lower U contents and higher ARs values at the more reduced zones of the aquifer. Despite these inverse relationships involving the U contents and ARs data have been verified in some aquifer systems (Cowart and Osmond, 1977; Toulhoat and Beaucaire, 1991), laboratory time-scale experiments conducted on limestone and dolomite gravels (Bonotto and Andrews, 2000) did not confirm these premises, since it was possible to obtain an inverse relationship between the AR and dissolved U in a reducing solution without considering any redox boundaries.

The effect of the temperature on the U speciation in Guarani aquifer was recognized (Bonotto, 1999a), where it was found that the bicarbonate and carbonate ions are important for the transport of U in solution. The data separation according to their negative and positive SI-calcite values allow identify interesting features concerning to the pattern exhibited by the values of AR and dissolved U content. Figs. 15b, 15c and 15i show that more oxidizing and acid conditions at lower temperatures favors the removal of the more labile ^{234}U isotope, implying on the generation of higher ARs values. Because uranium is expected to be released mainly from the sandstone matrix, lower values in solution (Figure 15h) are obtained when the SI-amorphous silica (Figure 15g) tend to reach the equilibrium value at lower temperatures.

Therefore, an inverse relationship is found between AR and dissolved U content (Figure 15j), but such possible requirement for the use of the U-isotopes chronological modeling is not feasible since the SI-calcite values (Figure 15f) do not suggest the occurrence of U-precipitation in the aquifer, a process necessary to enhance the AR by alpha-recoil.

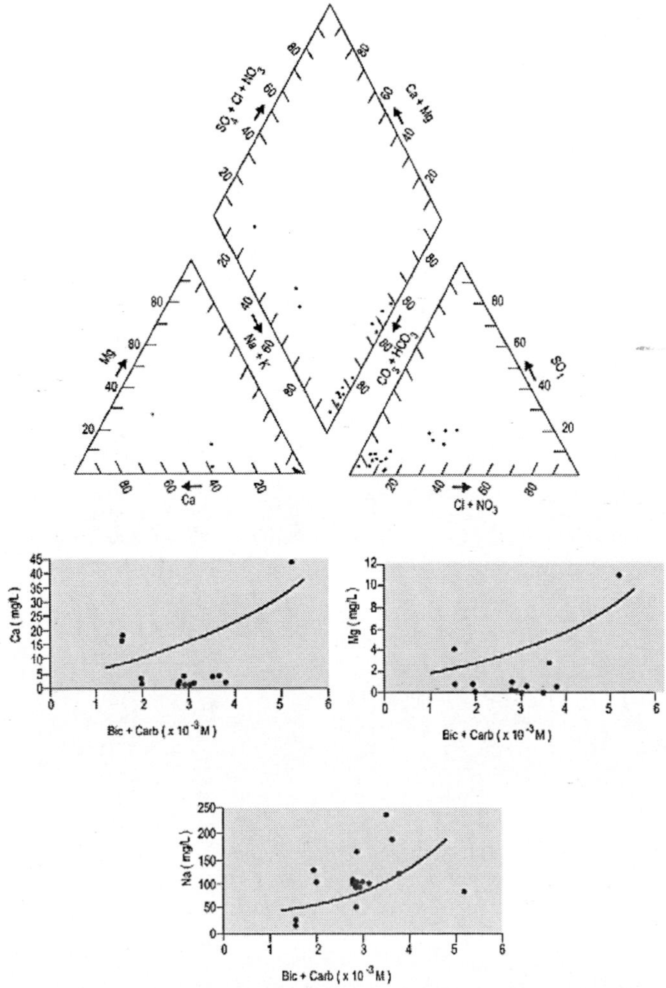

Figure 14. Chemical data for groundwaters from the Guarani aquifer having calculated positive SI-calcite values plotted on a Piper (1944) diagram and the amount of Ca, Mg, and Na dissolved in relation to bicarbonate + carbonate dissolution.

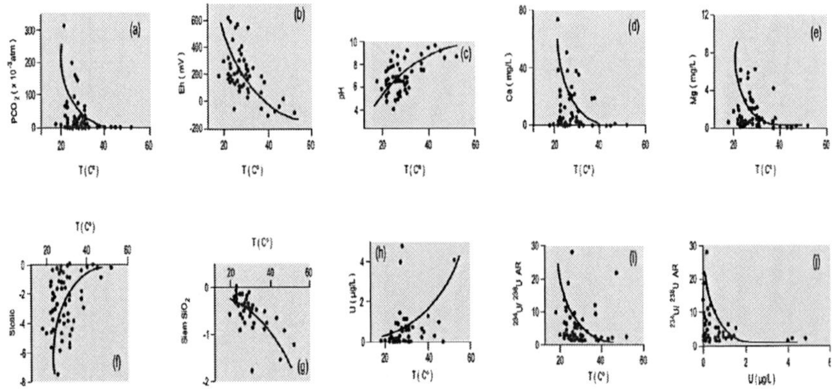

Figure 15. The (a-i) physical, chemical, mineral saturation indices and U-isotopes data plotted against the temperature and the (j) dissolved U content-$^{234}U/^{238}U$ activity ratio relationship for groundwaters from the Guarani aquifer having calculated negative SI-calcite values.

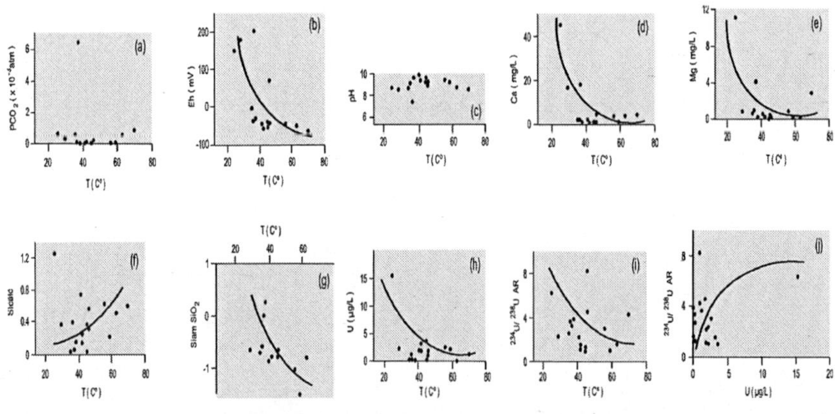

Figure 16. The (a-i) physical, chemical, mineral saturation indices and U-isotopes data plotted against the temperature and the (j) dissolved U content-$^{234}U/^{238}U$ activity ratio relationship for groundwaters from the Guarani aquifer having calculated positive SI-calcite values.

The data plotted in Figure 14 suggest that both uranium isotopes ^{238}U and ^{234}U precipitate with Ca (Figure 16d) and Mg (Figure 16e) with increasing temperatures in the aquifer at more reducing conditions (Figure 16b). The precipitation is more pronounced at higher SI-calcite values (Figure 16f), implying on the generation of lower AR (Figure 16i) and dissolved U content (Figure 16h) values, and also of a direct relationship between these parameters (Figure 16j). Again the U-isotopes chronological modeling cannot be utilized, because higher ARs values (Figure 16i) occur at the more oxidizing conditions (Figure 16b) in the aquifer. A combination of both data set as well is not plausible since there is no evidence of increasing AR at more reducing conditions in the two situations.

Therefore, the observed trends do not allow to use the U-isotopes chronological modeling and a possible explanation for this could be that the system is not enough closed, allowing the mixture of groundwater flowing through cracks, faults and joints of the basaltic package overlying this aquifer, among others factors. Under such circumstance, the uranium isotopes may be utilized for evaluating mixing processes (Osmond et al., 1974; Rosenthal and Kronfeld, 1982) rather than ages.

Chapter 5

CONCLUSION

Conventional hydraulic conductivity tests are important tools to investigate the groundwater movement and a wide range of them are available for generating the data either in the field or in the laboratorial context. Some experiments were described in this book, focusing the sand mining activities taking place at Analândia municipality, São Paulo State, Brazil. It was possible verify that despite of the sandy and porous characters of the soil covers in the area, the hydraulic conductivity (from 4.7 to 8.7×10^{-5} cm/s) and infiltration coefficients values (between 11.7 and 22.3 L/h/m^2) suggest a low susceptibility to the occurrence of this process that would imply on a "bad" category, according to typical classification systems. The low vulnerability was also evidenced by chemical analysis realized during others assays held in the area, i.e. solubilization and lixiviation, as well by periodical hydrogeochemical monitoring. The data of these monitoring programs for ground and surface waters were compared with maximum values established by reference guideline values proposed by the brazilian legislation and it was no verified the occurrence of any environmental impact related to the sand mining activities conducted in the area. The brazilian portion of the Guarani aquifer has been extensively investigated by many researchers and some data are here reported. Despite the U-isotopes modeling has been successfully applied in the aquifer for providing information on mixing studies and on the evaluation of the groundwater residence time, the approach here described tell us that some trends do not allow to use the U-isotopes chronological modeling and a possible explanation for this could be that the system is not enough closed, allowing the mixture of groundwater flowing through cracks, faults and joints of the basaltic package overlying this aquifer, among others factors.

ACKNOWLEDGEMENTS

FAPESP (Foundation Supporting Research in São Paulo State) and CNPq (National Council for Scientific and Technologic Development) in Brazil are greatly thanked for financial support of the investigations described in this book.

REFERENCES

ABGE (Associação Brasileira de Geologia de Engenharia) Ensaios de permeabilidade em solos: orientações para sua execução no campo; *Bol.* 4; ABGE: *São Paulo*, SP, 1981.

Andrews, JN; Kay, RLF. *U. S. Geol. Surv. Open-File Rep,* 1978, 78-701, 11-13.

Andrews, JN; Kay, RLF. *Isot Geosci*, 1983, 1, 101-117.

Andrews, JN; Giles, IS; Kay, RLF; Lee, DJ; Osmond, JK; Cowart, JB; Fritz, P; Barker, JF; Gale, J. *Geochim. Cosmochim. Acta*, 1982, 46, 1533-1543.

Araújo, LM; França, AB; Potter, PE. Giant Mercosul aquifer of Brazil, Argentina, Uruguay and Paraguay: hydrogeologic maps of Botucatu, Pirambóia, Rosário do Sul, Buena Vista, Misiones and Tacuarembó Formations; *Universidade Federal do Paraná: Curitiba*, 1995.

Baumgartner, A; Reichel, E. The world water balance; Elsevier: *Amsterdam*, 1975, 179.

Berner, EK; Berner, RA. The global water cycle; Prentice-Hall: Upper Saddle River, NJ, 1987, 397.

Bigeleisen, J; Wolfsberg, M. *Adv. Chem. Phys*, 1958, 1, 15-76.

Bonotto, DM. In *Water Pollution V. Modelling, Measuring and Prediction; Anagnostopoulos*, P; Brebbia, C. A; Eds; WIT Press: Southampton, 1999a, 281-292.

Bonotto, DM. In Proc. 3rd International Symposium on Environmental Geochemistry in Tropical Countries; SBGq (Sociedade Brasileira de Geoquímica); Ed; SBGq: Nova Friburgo, 1999b, CDROM.

Bonotto, DM; Andrews, JN. *Appl. Radiat. Isot.* 2000, 52, 965-983.

Borghetti, NR; Borghetti, JR; Rosa Filho, EF. da Aqüífero Guarani: a verdadeira integração dos países do Mercosul; Maxigráfica: *Curitiba*, 2004.

Bowie, SHU; Plant, JA. In Applied Environmental Geochemistry; Thornton, I; Ed; *Academic Press: London*, 1983, 481-494.

Castany, G. *Traité pratique des eaux souterraines*; Dunod: Paris, 1967.

Castany, G. *Principes et méthodes de l'hydrogéologie*; Dunod: Paris, 1982, 238.

Cowart, JB; Osmond, JK. *J. Geochem. Explor.* 1977, 8, 365-379.

Currie, LA. *Anal. Chem.* 1968, 40, 586-593.

Darcy, H. *Les fontaines publiques de la ville de Dijon*; V. Dalmont: Paris, 1856, 647.

de Marsily, G. *Quantitative hydrogeology: groundwater hydrology for engineers*; Academic Press, Inc: New York, 1986, 440.

Esteves, FA. *Fundamentos de limnologia*; Interciência: Rio de Janeiro, 1988.

Faure, G. *Principles of Isotope Geology*; Wiley: New York, 1986, 589.

Faure, G. *Principles and applications of inorganic geochemistry*; Mac Millan Publishing Co: New York, 1991, 626.

Fritz, P; Fontes, JC. *Handbook of Environmental Isotope Geochemistry;* Elsevier: Amsterdam, 1980.

Garrels, RM; Christ, CL. Solutions, minerals, and equilibria; *Harper & Row*. New York, 1965, 450.

Gilboa, Y; Mero, F; Mariano, IB. *J. Hydrol.* 1976, 29, 165-179.

IG (Instituto Geológico) Formações geológicas de superfície - Folha Geológica de Corumbataí - Fl. SF-23-Y-A-12; Instituto Geológico: São Paulo, SP, 1984.

IPT (Instituto de Pesquisas Tecnológicas) Mapa geológico do Estado de São Paulo; Monografias; IPT (Instituto de Pesquisas Tecnológicas): São Paulo, SP, 1981.

Ivanovich, M; Harmon, RS. *Uranium series disequilibrium: applications to environmental problems*; Clarendon Press: Oxford, 1982.

Kigoshi, K. Science, 1971, 173, 47-48.

Kimmelmann e Silva, AA; Rebouças, AC; Santiago, MMF. *Radiocarbon*, 1989, 31, 926-933.

Langmuir, D. *Geochim. Cosmochim. Acta.* 1978, 42, 547-569.

Lecomte, P. *J. Geochem. Explor.* 1988, 30, 35-61.

Liazi, A; Conejo, JGL; Palos, JCF; Cintra, PS. Rev *Águas e Energia Elétrica*, 1988, 14(5), 4-10.

Mook, WG. E*nvironmental isotopes in the hydrological cycle - Principles and applications; Technical Documents in Hydrology*; International Hydrological Programme, IHP-V, 39, UNESCO-IAEA: Paris-Vienna, 2000.

Mook, WG; Waterbolk, HT. *Handbooks for Archaeologists*; European Science Found: Strasbourg, 1985.

Oliveira, EG. Hidrogeoquímica aplicada na avaliação do impacto ambiental em áreas de lavra de areia; PhD Thesis; UNESP: *Rio Claro*, 2000.

Oliveira, EG; Bonotto, DM. In Anais XII Congresso Brasileiro de Águas Subterrâneas; ABAS (Associação Brasileira de Águas Subterrâneas); Ed; ABAS: *Florianópolis*, 2002, CDROM.

Osmond, JK; Cowart, JB. *At Energy Rev.* 1976, 14, 621-679.

Osmond, JK; Kaufman, MI; Cowart, JB. *Geochim. Cosmochim. Acta.* 1974, 38, 1083-1100.

Pinto, NLS; Holtz, ACT; Martins, JA. *Hidrologia de superfície*; Edgard Blücher: São Paulo, 1973.

Piper, AM. *Trans. Am. Geophysical Union.* 1944, 25, 914-928.

Raymond, LS. Jr. Groundwater Contamination; New York State Water Resources Institute, Cornell University: New York, 1988.

Rebouças, AC. *Episodes.* 1988, 11, 209-214.

Renne, PR; Ernesto, M; Pacca, IG; Coe, RS; Glen, J. M; Prévot, M; Perrin, *M. Science*, 1992, 258, 975-979.

Rosenthal, A; Kronfeld, J. *J. Hydrol.* 1982, 58, 149-158.

Rosholt, JN; Shields, WR; Garner, EL. *Science.* 1963, 139, 224-226.

Silva, RBG. Estudo hidroquímico e isotópico das águas subterrâneas do aqüífero Botucatu no Estado de São Paulo; PhD Thesis; USP: *São Paulo*, 1983.

Sitte, K. Handbuch der Physik 46; Springer Verlag: *Berlin*, 1967.

Tandel, RY. Caracterização do arenito Pirambóia da Fazenda São João em Analândia, SP, e sua utilização industrial; *Ms Dissertation; USP: São Paulo*, 1993, 75.

Teissedre, JM; Barner, U. Rev *Águas Subterrâneas.* 1981, 4, 85-95.

Todd, DK. Ground Water Hydrology; John Wiley & Sons, Inc: New York, 1959, 320.

Toulhoat, P; Beaucaire, C. *J. Geochem. Explor.* 1991, 41, 181-196.

Truesdell, AH; Jones, BF. *Jour. Res. U.S. Geol. Survey.* 1974, 2, 233-248.

Vries, JJ. de In Environmental isotopes in the hydrological cycle; In: WG; Mook, Ed; UNESCO-IAEA: Paris-Vienna, 2000; *International*

Hydrological Programme, IHP-V, Technical Documents in Hydrology, 39(1).

WHO (World Health Organization) 2004. *Guidelines for drinking water quality: radiological aspects* [http:www.who.int/water_sanitation_health/dwq/gdwq3/en/].

INDEX

A

B

C

U

V

W

X